Bio-Imperialism

Bio-Imperialism

～

*Disease, Terror, and the
Construction of National Fragility*

GWEN SHUNI D'ARCANGELIS

Rutgers University Press

New Brunswick, Camden, and Newark, New Jersey, and London

Library of Congress Cataloging-in-Publication Data
Names: D'Arcangelis, Gwen Shuni, author.
Title: Bio-imperialism : disease, terror, and the construction of national
fragility / Gwen Shuni D'Arcangelis.
Description: New Brunswick : Rutgers University Press, [2020] |
Includes bibliographical references and index.
Identifiers: LCCN 2020009740 | ISBN 9781978814783 (paperback) |
ISBN 9781978814790 (hardcover) | ISBN 9781978814806 (mobi) |
ISBN 9781978814813 (pdf) | ISBN 9781978815162 (epub)
Subjects: LCSH: Bioterrorism—United States. | National security—
United States. | Biological warfare—United States. | Racism—Political aspects—
United States. | United States—Foreign relations.
Classification: LCC HV6433.35 .D38 2020 | DDC 363.325/30973—dc23
LC record available at https://lccn.loc.gov/2020009740

A British Cataloging-in-Publication record for this book is available
from the British Library.

♾ The paper used in this publication meets the requirements of the
American National Standard for Information Sciences—Permanence
of Paper for Printed Library Materials, ANSI Z39.48-1992.

www.rutgersuniversitypress.org

Manufactured in the United States of America

For my Mom, Waipo,
and Waigong

Contents

Illustrations

Bio-Imperialism

Introduction

Bio-Imperialism and the Entanglement of
Bioscience, Public Health, and National Security

Not more than a month after the events of September 11, 2001, letters laced with deadly anthrax spores arrived at the offices of several news media outlets and two U.S. senators, causing five deaths and seventeen injuries.[1] The "anthrax attacks" swiftly shifted into position as a central node of the burgeoning war on terror, elevating its focus on bioterrorism, that is, the intentional spread of disease via germ or biological weapons. The FBI launched a massive, broad-scale investigation to find the perpetrator, enlisting advisers from science, national security, and policy and scholarly bodies, and even investigative journalists.

Early speculation reflected dominant national security discourse: the perpetrator was thought to be Al Qaeda or Iraq—the "usual suspects" of the post-9/11 era—or possibly Russia, a remnant of earlier Cold War era antagonism. Former vice president Dick Cheney conjectured that Al Qaeda "actually used to train people" in how "to deploy and use these kinds of substances [biological and chemical weapons]" ("The Anthrax Source" 2001). Weapons inspector Dick Spertzel was quoted in a *Wall Street Journal* article, "The Anthrax War," suggesting that "there are people in Iraq who know how to do it" (*Wall Street Journal* 2001b). There was little evidence for these theories.[2]

It took the FBI only a few months to determine that the anthrax used was the potent Ames strain, which derived from a U.S. government biodefense lab (USAMRIID).[3] By early 2002, the FBI had turned to investigating the high-level scientists who research pathogens for the U.S. national security apparatus, eventually identifying its final suspect, U.S. white male scientist Bruce E. Ivins.[4]

The investigation's findings were consistent with bioterrorism's historical pattern in the United States as a phenomenon carried out almost exclusively by domestic sources, specifically, white men.[5] While the federal government did not commission any political or other social scientific studies of white male violence, it continued to invest vast resources in soliciting scholars to study Arab and Muslim cultures as part of the war on terror.[6] (In fact, profilers in the anthrax investigation continued to focus on Al Qaeda and Iraq even after evidence pointed toward involvement by a U.S. government lab scientist.)[7] The lack of scrutiny of white male violence was consonant with dominant ideas about white masculinity as sources of authority and protection, ideas particularly entrenched during the war on terror (Shepherd 2006).

As has been theorized by cultural studies scholars, discourse—or ways of speaking about the world—constrains and limits the way that knowledge is constructed. Cultural theorist Stuart Hall defines discourse as "a group of statements which provide a language for talking about—i.e. a way of representing—a particular kind of knowledge about a topic. When statements about a topic are made within a particular discourse, the discourse makes it possible to construct the topic in a certain way. It also limits the other ways in which the topic can be constructed" (Hall 1992, 291). Dominant discourse marginalizes explanations—in this case the culpability of white masculinity—that do not fit its dictates.

Discourse is connected to power: knowledge production promotes some power arrangements over others. Again I cite Hall, who draws on social theorist Michel Foucault to explain the connection between thought and action, between language and practice: "Discourse is about the production of knowledge through language. But it is itself produced by a practice: 'discursive practice'—the practice of producing meaning. Since all social practices entail meaning, all practices have a discursive aspect. So discourse enters into and influences all social practices" (Hall 1992, 291). Discourse, then, has material effects; it is in fact formative of the material. Failure to problematize white masculinity in the anthrax case supported existing arrangements keeping white men in power—in the high-level U.S. biodefense industry, and at the helm of the U.S. national security apparatus more generally.[8]

Discourse surrounding the anthrax mailings not only marginalized the culpability of white masculinity; it also carefully bounded discussions about the vast research industry from which the anthrax came. The biodefense research apparatus spanned government, university, and industry labs, where bioscientists, mainly microbiologists, toil away on dangerous pathogens—in the name of national security. Over the course of the nearly nine-year anthrax investigation, the mass news media brought attention to the industry's ongoing lab accidents—accidental exposures, lab leakages, and unintentional shipments of live germs instead of dead ones. Media attention peaked in 2006, when several lab workers at Texas A&M University were exposed to and infected with Q fever and brucellosis, resulting in the lab's temporary suspension of activities. In response, federal officials insisted that the gains for national security outweighed the costs; security pundits concentrated on better training for scientists to reduce accidents. Absent among the proposed solutions, however, was a deeper questioning of the merits of the industry itself: What purpose does an active U.S. biodefense program serve?

Bio-Imperialism seeks to unravel the discursive edifices of U.S. biodefense: assumptions about bioterrorism and U.S. vulnerability, about germs and technoscientific capabilities to control them. It examines, moreover, the constitutive role that gender and race, along with U.S. imperial ambitions, play in U.S. bioterror and disease response.

The focus on Al Qaeda and Iraq during the anthrax investigation highlights the role of the racial Other in dominant national security discourse. The end of the Cold War saw the U.S.-USSR axis, and the concomitant capitalism-communism binary,[9] give way to the geopolitical and military supremacy of the United States, and the unprecedented acceleration of neoliberal capitalism and U.S.-led globalization (Masco 1999). The new lone superpower shifted its attention from the Soviet Union and nuclear stalemate to an increasingly visible number of smaller enemies—"terrorists"[10] and "rogue states"—and their possible acquisition of weapons of mass destruction (WMDs), which include weapons of chemical, biological, radiological, and nuclear warfare. Foreswearing de-escalation and disarmament, the United States turned to the platform of counterterrorism, most notably intervening in the Middle East.

U.S. involvement in the Middle East resulted in the targeting of Arabs and Muslims in the region as well as domestically. The United States passed the Anti-Terrorism and Effective Death Penalty Act of 1996 (AEDPA) in the aftermath of the Oklahoma City bombing perpetrated by U.S. white male Timothy McVeigh and two others, resulting in the deaths of 168 people and injuring over 800 in April 1995. Despite the domestic origins of the perpetrators, AEDPA focused heavily on "international terrorism" and "alien terrorists," and in practice was used to apprehend Arabs and Muslims. It authorized the secretary of state to designate "foreign terrorist organizations" that the United States could sanction. About half were Muslim or Arab groups (Whidden 2001).[11] As Arab American studies scholars have demonstrated, the U.S. state justified these actions by mobilizing trenchant U.S. racial discourses villainizing Arabs and Muslims as uncivilized, violent Others[12] (Cainkar 2008; Naber 2000).

By the time of the September 11 incident, the specter of Arab/Muslim[13] terror had been firmly entrenched, serving as the lynchpin for a new iteration of U.S. empire. The war on terror entailed the brutal invasions of Afghanistan and Iraq to maintain U.S. oil interests and a foothold in the Middle East.[14] In President George W. Bush's address to a joint session of Congress on September 20, 2001, he called for military action that would begin with Al Qaeda and continue "until every terrorist group of global reach has been found, stopped and defeated" (Bush 2001a). The war's boundless scope was in part enabled through the larger shift in governmental power during the war on terror—what social theorist Brian Massumi (2010) has called "preemptive power." This anticipatory mode of governance enhanced the U.S. state's ability to act upon security threats of indeterminate potential in the name of preserving life.

Preemptive state power was, moreover, yoked to the language of freedom and civilization. In his September 20 speech, Bush justified military action as necessary to "defend freedom" and invoked "civilization" as the war's protagonist: "This is civilization's fight" (Bush 2001a). Narratives of the United States as freedom-loving, democratic, just, and other markers of "civilized" can be traced to long-standing Orientalist discourse that positions the West as beacon of progress and opposes it to the Arab/Muslim Other. These discursive practices began in the late eighteenth century, when European narratives of the "East" extolled the superiority of the "West" in order to rationalize colonial endeavors in the Middle East (Said 1978).

Critical race and ethnic studies scholar Sylvia Chan-Malik discusses how this Orientalist narrative has functioned in relation to U.S. aggression in the Middle East: as "the liberal vision of a free, feminist, and multicultural nation as a fundamental necessary counterpart to the decidedly unfree, antifeminist, and antidemocratic ideology of Islamic Terror" (2011, 134). In other words, the construction of Arabs and Muslims as a racialized enemy Other helped inculcate a national imaginary of the United States as progressive vis-à-vis feminism and multiculturalism.[15] Bush's speech on September 20 included numerous nods to multicultural tolerance, such as "The enemy of America is not our many Muslim friends; it is not our many Arab friends" (Bush 2001a). This rhetoric belied the fact that the war clearly targeted Arabs, Muslims, and the Middle East.

Bush's speech also touched on the plight of women in Afghanistan: "Women are not allowed to attend school" (Bush 2001a). Arab American feminist studies scholars have thoroughly described the construction of Arab and Muslim women as always and everywhere oppressed, wherein regional distinctions are flattened, and women are seen solely as victims of a patriarchal culture. These scholars have demonstrated how this image has propped up the war on terror—the war's proponents articulate the war as an attempt to liberate oppressed Arab and Muslim women (Moallem 2002; Nayak 2006). In fact, Chan-Malik's discussion of the role feminism has played in exceptionalist constructions of the United States illustrates a dialectic in which the oppressed Arab/Muslim woman is the counterpoint to the liberated Western woman, further bolstering an image of U.S. progressivism despite U.S. imperial aggression.

Gender has in fact served as a nimble tool in the narratives of the war on terror. U.S. political theorist Alyson Cole (2007) describes, for example, the metaphor of sexual conquest that many public commentators used to describe the September 11 attacks: September 11 as the day the United States lost its virginity, or the attacks on the United States as akin to sexual assault. The metaphor of womanhood reinforced the notion of the United States as innocent victim of the violent masculine Arab/Muslim Other. Queer theorist Jasbir Puar dissects U.S. popular culture depictions of Osama bin Laden that pathologized him through the theme of deviant sexuality—images of bin Laden being penetrated anally or descriptions of him as possessing a kind of queer vanity (2007a, 38). These queer overtones in the portrayals of bin Laden served to construct him as what Puar calls "monster-terrorist-fag," the ultimate reviled, dangerous Other to justify the war on terror.

In the war's turn to bioterrorism, popular narratives similarly mobilized gendered discourses of an Arab/Muslim Other. Alongside investigator focus on Al Qaeda and Iraq during the anthrax case, mass culture producers depicted the anthrax perpetrator as a crazed, violent Arab/Muslim male—even though this image was totally unmoored from any basis in a real person or suspect. In chapter 1, I examine more closely the trope of the masculine Arab/Muslim Other in discourses on bioterrorism; further, I explore a notable departure from the ubiquitous trope of the oppressed Arab/Muslim woman: U.S. depictions of Iraqi female scientists apprehended by the United States during the Iraq War as devious and dangerous Arab/Muslim female terrorists.

The Bioterror Imaginary Takes Hold

The anthrax investigation and focus on bioterrorism were not just a window into the dominant discourse on terrorism, but also on disease. Media coverage of bioterrorism skyrocketed during the investigation,[16] and many articles contained elaborate scenarios of deadly germs coupled with the specter of malevolent violence. One such article described "easily obtainable lethal chemicals and viruses" and an "emerging picture of budding do-it-yourself biological unibombers" (*Greensboro News and Record* 2002). The depiction of germs as easily accessible weapons conjured an image of bioterrorism as a growing phenomenon.

To this unsettling picture, government and news media added the notion that the United States was unprepared. In the aftermath of September 11, in fact even before the mailings were discovered, the *New York Times* had inaugurated a new section, "The Biological Threat," in its post-9/11 series "A Nation Challenged."[17] In one article in the series, "Nation's Civil Defense Could Prove to Be Inadequate against a Germ or Toxic Attack," published on September 23, 2001, authors William J. Broad and Melody Petersen painted this portrait: "Experts say civil defenses across the nation are a rudimentary patchwork that could prove inadequate for what might lie ahead, especially lethal germs, which are considered some of the most dangerous weapons of mass destruction" (2001). The article reflected a common view among military and scientific communities that germ/biological weapons are potent but unpredictable entities that cannot be entirely controlled once unleashed (Cecire 2009, 47).

What I call a "bioterror imaginary" emerged, comprising a landscape of ideas, meanings, sensibilities, and subjectivities centered on the threat of bioterrorism and U.S. vulnerability. This imaginary, moreover, extended beyond government and news media domains. TV shows, films, novels, and other forms of entertainment took on the subject of bioterrorism, as did individuals who published do-it-yourself response manuals, books, websites, and other very publicly accessible products. An episode of the TV military drama *E-Ring* titled "Breath of Allah" (2006) focused its plot on the agents' discovery of a lab producing plague in Amsterdam—the agents subsequently look for cities with large Muslim populations and mosques. Laura Landro, author of the *Wall Street Journal* health column "The Informed Patient," offered her advice: in "Don't Leave It All to Doctors to Know Signs of Bioweapons," she urged readers to "spend a little time yourself getting educated about the risks, symptoms, and treatments" as well as to look on the Internet for "a wealth of information about anthrax, smallpox and other threats such as tularemia that consumers can easily understand" (Landro 2002).[18]

Whether directly referencing an Arab/Muslim terrorist or not, these scenarios strengthened this feature of bioterrorism discourse. Meaning accumulates across mass media texts; the words, sounds, and images contained in one refer to others, their meaning altered by being read in the context of one another, a process that Stuart Hall articulates as "inter-textuality" (1997, 232). The reach of mass culture realms is, moreover, a key facet of its ability to entrench discourse, in this case the assemblage of germs, disease, and Arab/Muslim terror. The fact that the culprit of the anthrax mailings—the pivotal post-9/11 bioterror incident in the United States—was not in fact the maligned Arab/Muslim terrorist, did nothing to stem the tide of this racialized imaginary.

As the bioterror imaginary took root, government officials, with the help of pundits from the national security and bioscience fields, formulated a regime of bioterrorism preparedness. Prominent biodefense pundits Tara O'Toole and Thomas Inglesby outlined three focal areas in the inaugural issue of the peer-reviewed journal *Biosecurity and Bioterrorism: Biodefense Strategy, Practice, and Science*:[19] biodefense research and development, medical and public health capacity, and prevention of bioweapons development and use (2004, 2). The first, the revamping of the biodefense industry, tackled the system of high-level labs that Ivins was involved in, where scientists conduct research and development on lethal pathogens and their countermeasures (e.g., diagnostic tests and vaccines). From fiscal years 2002 to 2004, the National Institute of Allergy and Infectious Diseases (NIAID), the government agency responsible for conducting infectious disease research, had increased its budget by more than twentyfold for research into anthrax, smallpox, plague, and other lethal pathogens used in biological warfare (Sunshine Project 2005).

This research on lethal pathogens was driven not only by the bioterror imaginary, but also by a faith in the ability of technoscience to yield solutions to social problems.[20] This faith has its origins in Euro-American modernity and its centering of science and technology as vehicles of cultural progress and advancement (Foucault 1977). It has meant the promotion of innovation and the prioritization of technological solutions over social ones. In the context of biodefense, it has entailed pouring resources into, for example, new vaccine development and "threat characterization" studies. The latter entail research involving the production of lethal pathogens—sometimes new variants such as the anthrax Ivins worked with—so that new vaccines may be tested against them, or simply to gain knowledge about possible new biological weapons that enemies of the United States might produce.

While many bioweapons specialists were supportive of the lean into technoscientific innovation in germ research, bioweapons specialist Jonathan B. Tucker voiced a cautionary note on this unbounded germ research, noting that "creating putative bioengineered pathogens in the laboratory for purposes of threat characterization would be exceedingly dangerous and counterproductive" and that "there is an important distinction between 'knowledge gaps' that are worth filling and those whose exploration could generate new dangers" (Tucker 2006, 195–196). For Tucker, the production of dangerous new pathogens had myriad issues whose blowback would inevitably harm the United States: the danger posed to scientists via accidental exposure, the acquisition of U.S. weapons by its enemies, and possible negative perception that U.S. creation of biological weapons for research was a ruse for the creation of biological weapons for warfare (Tucker 2006). Tucker's critique highlights that the Bush administration's pursuit of innovation in the name of bioterrorism preparedness, and its expectation of future benefit deriving from technoscientific advancements, obscures the significant dangers such innovation produces. In chapter 2, I discuss the way the dangers of germ research were sidelined vis-à-vis discourses of the racial Other encoded in both legal measures (the Bioterrorism Preparedness and Response section of the PATRIOT Act and the new Bioterrorism Preparedness Act of 2002) and "biosecurity" practices instituted in the biosciences.

A Short History of Biological Warfare

Biodefense research has a longer history that precedes its post-9/11 makeover. Biodefense research on lethal pathogens was the legacy of earlier regimes of biological warfare. European settler-colonists during the 1700s waged the very first intentional deployment of lethal pathogens in the United States—spreading smallpox to indigenous populations, decimating them (Christopher et al. 1997; Duffy 2002; USAMRIID 2004). Following its role in the genesis of U.S. settler-empire, biological warfare played an important role in U.S. military programs during the early- to mid-twentieth century, when many nations developed large-scale programs[21] to weaponize a variety of germs (i.e., bacteria, viruses, fungi, protozoa) and other biological weapons[22] (Barnaby 2000; Bernstein 1987; Clarke 1968; Hersh 1968).

After World War II, U.S. engagement with biological weapons receded in light of the rise of nuclear weapons as well as the global decline of state biological weapons programs. In 1972, an international treaty, the Biological Weapons Convention (BWC), banned the development, possession, and transfer of biological weapons. Only scaled-down "defensive" programs remained. These programs ebbed and flowed over subsequent presidencies, until the late 1990s, when they received a substantial boost.

The Clinton administration developed a two-pronged program. On one end, the administration focused on domestic preparedness, emphasizing broad technological solutions and emergency response, establishing the National Pharmaceutical Stockpile and the Health Alert Network and Laboratory Response Network in 1999 (Guillemin 2005a; Khan, Morse, and Lillibridge 2000). On the other end, the administration monitored enemy nations such as the former Soviet Union as well as nonstate groups and individuals the United States deemed terrorists. The United States surveilled, for example, the biological warfare activities of nonstate groups such as Aum Shinrikyo[23] and investigated whether the dissolved former Soviet Union[24] had transferred any biological weapons to states in the Middle East, such as Iraq[25] (Fidler 2002; Mueller 2005; O'Toole 2001).

The attention to biological weapons stemmed in part from U.S. post–Cold War focus on terrorism and efforts to secure its global hegemony. Another basis of concern was the advances in biotechnology that had emerged since the birth of genomics in the 1980s, and in genetic engineering in the 1990s, which created the possibility of enhanced germ weaponry.[26] Since offensive biological weapons were banned, the U.S. government exerted dominance by developing biodefense and suppressing weapons capacity in others.

Clinton's global watchdog approach would become the foundation for post-9/11 engagement. Under the Bush administration, the United States invaded Iraq in 2003 with the rationale of preempting Iraqi use of bioweapons (and other WMDs). The administration offered no proof of either Iraqi weapons possession or plans to attack the United States; instead, it offered alarmist allusions—to, for example, Iraqi connections to Al Qaeda or the active Iraqi bioweapons program in the 1970s and 1980s (which had been destroyed during UN inspections in the 1990s). Bioweapons had become a powerful rhetorical tool in U.S. designs on Iraq and signaled that the United States would attempt to justify drastic military action in the name of countering bioweapons threats.

At the same time the Bush administration maintained a tight grip over the bioweapons capacities of others, it dramatically expanded Clinton's domestic preparedness program—in areas such as biodefense research, vaccine stockpiling, disease surveillance networks, and response planning infrastructure. On June 12, 2002, Bush outlined this far-reaching approach in his remarks accompanying the signing of the Public Health Security and Bioterrorism Preparedness and Response Act of 2002: "Bioterrorism is a real threat to our country. It's a threat to every nation that loves freedom. Terrorist groups seek biological weapons; we know some rogue states already have them. . . . It's important that we confront these real threats to our country and prepare for future emergencies" (Bush 2002c). The quote illustrates the key logic of preparedness—it assumes the inevitability of catastrophes like bioterrorism and opts to prepare for their aftermath (Ben Anderson 2010).[27] Moreover, the degree of resources devoted to preparing for a given catastrophe is determined not by the catastrophe's probability of occurrence, but by its potential magnitude; thus, threats viewed as low probability, high impact—like bioterrorism—garnered massive resources that would otherwise be devoted to more frequent, everyday concerns (Lakoff 2008b). Like preemption, preparedness has the effect of making extreme actions seem necessary, and other options, namely, complete disarmament and an end to biodefense research, seem inadequate. Preparedness is an anticipatory mode of governance that, as social theorist Melinda Cooper (2008) has described, is highly generative—it calls forth a future that "is effectively generated de novo out of our collective apprehensiveness" (125).

Preparedness, then, typified the profound feat of the war on terror—its adoption of the narrative of victimhood. To be prepared for outside threats presumed U.S. fragility,[28] rendering U.S. global aggression invisible. Centuries after European settlers used biological weapons against Native Americans, the United States—now a white-dominant imperialist nation with an augmented biowarfare capacity—articulated itself as at the mercy of Iraq, Al Qaeda, and other Arab/Muslim groups and nations.[29]

Health, Security, and Social Control

Bob Stevens was the first victim of the anthrax mailings. He was a photo editor for a media company in Florida—American Media Inc., one of the destinations for the anthrax-filled letters. On October 2, 2001, Stevens checked into a hospital. Infectious disease specialists examined a fluid sample taken from Stevens, first at the hospital, then later at one of the state's laboratories equipped to identify infectious agents potentially used as weapons, such as anthrax. The lab identified inhalation anthrax, which was extremely rare in the United States. Soon after, state health officials, the Centers for Disease Control and Prevention (CDC), law enforcement, and the FBI were called in. They held a press conference on October 4, announcing what would be the first of the 2001 anthrax cases. Sadly, Stevens died the next day. Over the course of the month, as more people emerged with anthrax infections and anthrax-laced letters were found, officials proclaimed that the anthrax was the result of a deliberate attack rather than a natural occurrence (Department of Justice 2010; L. Cole 2003).

The involvement of infectious disease specialists in identifying the anthrax illustrates a key role that the health field would come to play in post-9/11 bioterrorism preparedness. Health workers were charged with acting as sentinels for bioterror: preparedness programs trained them to recognize the symptoms of bioterrorism diseases such as anthrax and to report outbreaks that might be the result of such diseases to law enforcement. Preparedness efforts also employed disease surveillance and response infrastructure for the detection and communication of potential bioterrorism-caused outbreaks. In turn, the national security field gained greater influence over the health field. In 2002, the newly formed Department of

Homeland Security (DHS) absorbed various federal departments and/or their functions—including health-related ones—under a militarized mandate to act in the name of national security.[30]

The mobilization of public health for national security purposes was not without its critics. Health practitioner-researchers Victor W. Sidel, Robert M. Gould, and Hillel W. Cohen questioned what they called the "cooptation of public health"; they explained how it "might compromise the independence of public health professionals and agencies and subordinate their priorities to the priorities of the military, intelligence, and law enforcement agencies themselves" (2002, 86). They highlighted the detrimental effects on public health, such as the diversion of resources from more pressing chronic health problems, and suggested that the threat on which anti-bioterrorism programs were based was intentionally exaggerated to support military programs and national security agendas (Sidel, Gould, and Cohen 2002, 83–84). In chapter 3, I build on their critiques and analyze the gendered dimensions of the recruitment of public health into the war on terror. I focus on the mobilization of caregiving tropes and discourses of feminized vulnerability—namely, images of vulnerable white women—into bioterrorism preparedness.

The overlap of health and security objectives was not new to the post-9/11 period. The late 1800s saw the white majority and health authorities crack down harshly, and often preemptively, on racialized immigrant groups such as Chinese and Mexicans, whom they viewed as disease threats.[31] These disease control campaigns surveilled, criminalized, and quarantined marginalized groups (communities of color, female and gender-nonconforming communities, working-class populations) in order to protect the health of dominant groups (Kraut 1994; Leavitt 1996; Molina 2006; Shah 1999; Stern 1999; Tomes 2000). Hegemonic views of these groups as public health menaces were only in part about disease; they were also about the social threat such groups posed to elites, whether that be in terms of labor competition, population numbers, or rising social status. Public health, then, acted as a means to secure the dominance of already dominant groups (i.e., white, male, of means).

As U.S.-led globalization peaked toward the end of the twentieth century, the U.S. state focused heavily on the global spread of infectious disease. When HIV/AIDS emerged in the United States in the 1980s, it challenged the common U.S. public health view that infectious disease was a problem of the past. The disease soon spread globally; it became difficult to contain, and, moreover, endemic in many African regions. This led the Clinton administration to formally designate it as a threat to U.S. national security in April 2000: the administration linked the death and destruction caused by HIV/AIDS in the Global South to the creation of social instability in these regions, which it then viewed as spawning political regimes and groups hostile to the United States ("Clinton Administration" 2000; Gellman 2000). This framing of HIV/AIDS (and later other types of global epidemics) through the lens of security revisited the role of public health in maintaining U.S. power hierarchies—in this case U.S. global hegemony.

In the post-9/11 context, two diseases became the focal point of the U.S. disease-security imaginary: severe acute respiratory syndrome (SARS), which emerged in Guangdong, China, in late 2002; and H5N1 avian influenza ("bird flu"), which emerged in Hong Kong the following year. U.S. authorities invoked tropes historically applied to diseases associated with bodies of color—in this case Orientalist tropes of Asia and Asians as a dirty and unhygienic people.[32] They also met SARS with the anxieties of the post-9/11 era: U.S. pundits speculated about its potential origins as a germ weapon,[33] or alternatively that it could be made into one[34] (both have been considered unlikely by the majority of SARS scholars and researchers).

By the time H5N1 influenza (bird flu) emerged in Hong Kong at the end of 2003, terror rhetoric had definitively worked its way into disease control discussions. An Institute of Medicine (2003) report stated, "Influenza is an exemplar of nature's natural biowarfare." Here, "nature" is anthropomorphized and given ill intent associated with human acts of violence—that is, biowarfare.[35] As international concern grew that H5N1 would turn into a pandemic (a disease that spreads on a global scale), the United States devised preparedness measures aimed to address both flu and bioterrorism pathogens. In 2004, the Department of Homeland Security set up the National Biosurveillance Integration Center, which aimed to integrate all the health surveillance systems across the country into a single system, collecting information on both bioterror events and naturally arising disease trends.

This integrated approach informed the subsequent regime of pandemic preparedness, which focused on securing the United States from pandemics and potential pandemics arising in the Global South (such as SARS and H5N1 flu). In chapter 4, I examine the way that SARS and H5N1 influenza became the pivots of a security-inflected pandemic preparedness regime. I focus on how the United States mobilized a pandemic threat imaginary to justify extending its reach over global disease control—that is, U.S. ability to dictate the disease governance and resource allocations of nations and regions around the world.

Four Sites of Discourse-Making: Tracking Post-9/11 Bio-Imperialism

Imperialism more broadly denotes European practices (begun in the fifteenth century and inherited by the United States in the twentieth century) of global economic expansion, subjugation of colonized peoples around the world, and the ideologies of racial hierarchy and progress that bolstered them (Loomba 2015; Smith 2002).[36] *Bio*-imperialism signals the bureaucratic systems and accompanying ideologies that control and commercialize local biological materials such as plants and genetics for the benefit of global elites (Dorsey 2004). In the twenty-first century, the dominant countries of the Global North—and elite populations within these countries—advance their health, science, and military interests through this global rule over biological resources.[37]

This book tracks bio-imperialism vis-à-vis U.S. production, management, and distribution of a vital biological resource—germs. It focuses on the bioterrorism and pandemic preparedness regimes as the means by which the United States multiplied its germ resources between 2001 and 2008 (roughly). Further, this book tackles discourse-making as central to these preparedness regimes.

Cultural studies approaches aim to discern the conditions and means by which dominant discourses are made (and unmade). They focus on how individual actors reproduce, rework, or alternatively challenge discourses that are bound up with dominant ideologies and interests. I employ a cultural studies approach drawn from the fields of transnational feminist studies/postcolonial feminist studies/women of color studies, as well as ethnic studies and American studies—in particular scholarship within these fields engaged with science, technology, and medicine. These fields, seeking to pave the way for social justice and liberation, critically interrogate the U.S. state and dominant social groups (e.g., whites, males, upper-class groups) and the discourses they deploy. Following this well-trodden path, *Bio-Imperialism* analyzes multiple domains central to generating the specter of bioterrorism and disease, from popular culture realms (e.g., media) to specialized sites (e.g., science and health institutions) to state apparatuses (e.g., the Department of Homeland Security).[38]

Specifically, this study draws on documents from the federal government, news media, public health, and bioscience institutions. I analyze the power/knowledge formations across them, as sites of what Michel Foucault articulates as "governmentality": attempts at governing that problematize individual and collective conduct, attempts implemented not only by the state but also by other societal institutions (churches, hospitals, and prisons were some of the many institutions Foucault focused on).[39] Nancy Fraser elaborates practices of governmentality as "a distinctive set of regulatory mechanisms which suffused them [institutions] with a common ethos," and further states that "widely diffused throughout society, these small-scale techniques of coordination organized relations on the 'capillary' level: in factories and hospitals, in prisons and schools, in state welfare agencies and private households, in the formal associations of civil society and informal daily interaction" (Fraser 2003, 162). *Bio-Imperialism* considers the processes and practices across four locuses of governmentality as germane to understanding the power/knowledge formations involved in the preparedness regimes.

The federal government, as the formal locus of governance spearheading the preparedness regimes, constituted a primary focus. I analyzed key sites governing national security, science, and disease control (e.g., the Centers for Disease Control and Prevention or the Department of Homeland Security). These sites are constructed by and through discourse expressed in various texts such as laws, policies, and officials' rhetoric; I collected texts pertaining to the George W. Bush administration's approach to national security, science, and disease control. I accessed primary documents that were publicly available via government websites and congressional records. I also researched secondary sources such as legal and policy analysis scholarship (assisted by the online search tool Nexis Uni [formerly LexisNexis Academic]).

The second complex of sites I drew on entailed health care and biomedical research institutions—where bioterrorism- and disease-control policies and practices were implemented, negotiated, and reworked. I perused the guidelines and statements of major policy-making bodies and organizations such as science journal editorial boards and nursing associations tasked with bioterrorism preparedness. Their statements and actions reflected discursive and institutional shifts in response to national security measures. Relevant documents were collected from the institutions' websites and other sites of cultural production, such as newsletters and secondary government and news media sources.

The mass news media constituted a third research locus—both as a primary source of officials' statements and national security information otherwise inaccessible to the public, and as a vital meaning-making site where journalists produced and disseminated ideas about science, health, and national security. News media are characterized by a diversity of viewpoints—how journalists choose to convey their stories can vary. At the same time, this variability is constrained by media frames. Dominant discourses shape the types of stories that appear in the media and the way that new events get incorporated into existing frames. Thus, even multiple viewpoints can mobilize similar underlying messages.[40]

Media format (TV broadcast, social media, print news) also shapes message construction and circulation. I focused on corporate, mass news media because of their powerful influence over public news consumption: corporate monopolies possess the ability to homogenize media messages and, with the aid of communication technologies such as the Internet, propagate them widely (Poster 1995; Stewart, Lavelle, and Kowaltzke 2008). Corporate news media is thus a key site for the maintenance of dominant discourses. I examined five of the top seven most read newspapers (in 2007) in the United States according to BurrellesLuce: *USA Today*, the *Wall Street Journal*, the *New York Times*, the *Los Angeles Times*, and the *Washington Post*.[41] Main themes were identified through a progressive theoretical sampling of articles—with an eye trained on ideologies of gender, race, and nation; I used the search engine ProQuest News as well as Access World News[42] to survey the timeline of topics in U.S. news media as a whole—when there were more or fewer articles on a particular subject.

My fourth and final research locus entailed civic and social organizations expressly dedicated to working toward social justice and to addressing state violence. While I primarily analyzed the construction of the hegemonic discourse upholding U.S. (imperial) state power, I paid attention to the overall landscape of discourses and attended to instances when officials, scientists, health workers, and journalists produced counter-hegemonic discourse. Counter-hegemonic discourse, in my view, comprises ideas and ways of thinking that forward social justice and liberation perspectives and dismantle racism, Islamophobia, sexism, and other oppressive systems. Thus, I also researched public-oriented watchdog groups such as the ACLU and the Sunshine Project, both of which tracked U.S. biodefense research and provided critical perspectives and interventions on post-9/11 shifts in security, science, and health. I examined their reports and white papers not only for primary data on counter-hegemonic discourses (as well as noted whether and how hegemonic discourses were present even in social justice organizations), but also for key secondary data on the often secretive, less publicized actions taking place in the national security arena.

A Feminist Scholar's Journey: Curiosity, Accountability, Intervention

As a feminist attentive to the interplay of gender, race, and nation, I condemn the war on terror and the violence conducted in its name, as well as its intrusion into (ideally) nonmilitarized realms— from immigration and law enforcement to the science and health realms that are my focus. As someone who has lived much of her adult life during the war on terror, witnessing its targeting of Arabs, Muslims, and South Asians, as well as other communities of color, I am invested in dismantling the colonial narratives that further U.S. empire and harm women and communities of color transnationally. My interest and approach are informed by my own social position as a U.S. woman of color with lived experience as an Asian American and ties to East Asia specifically—I am sensitive to both gendered racial targeting and Orientalist narratives about Asia and Asians. The collective, critical perspectives of impacted groups that I both belong to and do not belong to drive my work.

The bio dimensions of the war on terror perhaps piqued my interest because of my background in biology. As a college student, I became interested in understanding the inner workings of humans, and pursued an interdisciplinary degree in neuroscience that integrated study of the biosciences with psychology. After college, I worked for five years as a laboratory technician, in sleep behavior and genetics laboratories, but came to realize that my true passion lies in inquiry at the social, rather than the biological, level. Eager to understand social structures, power, and inequality, I turned to graduate school—in women's studies. There, I delved deeper into feminist science studies, which bridged my interest in questions about science and society, particularly their gendered, racial, and other identity/power dimensions.[43]

It was a few years into the war on terror, in 2005, that I became both unsettled by and curious about a new development in the discourses on terrorism: a previously little-used phrase—"biological threats"—had cropped up in academia to denote germs, whether naturally arising or weaponized,[44] and was accompanied by the recruitment of academic researchers from science, medicine, and health to efforts to address bioterrorism. A flier circulated at UCLA, where I was located at the time, titled "Public Policy and Biological Threats, an IGCC Training Program"; it described a summer seminar sponsored by the University of California Institute on Global Conflict and Cooperation (IGCC) at UC San Diego on "policy responses to bioterrorism and emerging public health threats." Initiated in 2004, this institute brought together national security pundits, public health officials, epidemiologists, and other disease specialists.

Aware that anthropologists and religious studies scholars were being enlisted to support the war on terror, I suspected that the IGCC training program was of a similar ilk but recruited a new set of scholars—from the fields of biology and medicine. I decided to enroll in the three-week seminar, which went from July 18 to August 5, 2005. During my time there, I functioned as an insider-outsider, a sort of participant-observer able to witness firsthand the enlistment of various types of participants: some from security studies and political science—who generally possessed a neoconservative orientation; some from epidemiology and other health sciences—who seemed to show little interest in geopolitics but were vaguely supportive of the national defense agenda. The plethora of guest instructors reflected this array of fields—we were treated to lectures by high-level officials from the Clinton and Bush administrations, prominent bioweapons experts, epidemiologists, and pundits from the U.S. military–affiliated RAND Corporation, all with the firm conviction that "biological threats" merited a concerted effort.

This seminar proved a unique opportunity to identify the actors and stakes involved in larger mobilization efforts for bioterrorism preparedness. It provided a glimpse into the combination of neoconservative security politics with the technical know-how of the bioscience and health fields. When I returned from my brief visit to "the field" and shared my experiences with feminist colleagues, they met me with the distress I expected—the seminar seemed yet another example of the proliferating post-9/11 counterterrorism regime. But I was also met with some bewilderment about bioterrorism, the idea of which evoked the mystification surrounding science in general. This reflected a main finding of feminist science studies scholars: that outsiders to science (in this case my feminist colleagues) and scientists (my colleagues in the seminar) alike are subject to dominant narratives of science that render invisible its political dimensions. Exposing these dimensions, I have come to believe, is an important means to empower those outside of the science domain (and those within) to engage meaningfully in shaping the course of scientific practice, in this case to challenge the uncritical partnering of science with the U.S. national security apparatus.

I imagine this book will be of interest to critics of the war on terror and U.S. imperialism. *Bio-Imperialism* provides a moment of reflection—to consider and deconstruct state-led response regimes and the discourses that accompany them. It urges readers to question the way the U.S. state articulates threats—to examine the assumptions, stakes, and goals embedded in discourses that define particular groups and entities as "threats" and others as "vulnerable." *Bio-Imperialism* aims to provide readers with the means to scrutinize the state's enlistment of the biosciences and public health into bioterrorism/pandemic preparedness regimes.

My exploration of the nexus of national security, bioscience, and public health, through particular focus on gender, race, and nation, aims to center the impacts on and implications for targeted groups—to highlight how oppressive systems affect actual people. This includes Arabs, Muslims, South Asians, and others targeted by the U.S. national security apparatus, and East and Southeast Asians ensconced in U.S. disease control measures globally. It also includes bioscience researchers and health workers, especially those in targeted populations and regions. While my work concentrates on the workings of U.S. empire—to illuminate the sites and possibilities for intervention on hegemonic power—it also strives to account for the ways that targeted groups have engaged and dissented.

In particular, I position my work alongside—and in conversation with—Arab, Muslim, and anti-imperial feminist scholarship that both centers the impacts of the war on terror on Arab and Muslim communities and explores everyday issues and experiences for Arabs and Muslims beyond these oppressive systems. Feminist cultural anthropologist Nadine Naber (2011) highlights how, when feminists and other critical scholars write about the impacts of oppression on marginalized groups, they at times present them as only targets and victims, foregoing a discussion of *what matters to these groups on their terms.* Such writing practices unwittingly reproduce what they aim to diminish—namely, the marginalization of these groups. On the other end of this problem, scholarly attempts to center marginalized groups with detailed descriptions of their resistance tactics and everyday concerns can also reveal information that dominant actors can then utilize to further oppress these groups. The state's enlistment of U.S. scholars with expertise on Iraqi culture—during and on behalf of U.S. occupation—serves as one blatant example. Critical ethnic studies scholars John D. Márquez and Junaid Rana mark the distinction between efforts "to theorize and better understand oppression so as to encourage more effective methods to unsettle and disrupt it," and efforts "to gather and disseminate facts about oppressed peoples, a praxis that often contributes to oppression" (2015, 6).

I aim to navigate these twin dilemmas (linguistic objectification vs. overexposure) by representing targeted groups as subjects via using humanizing language and substantively centering their perspectives, but avoiding deep description of their resistance tactics. With these writing methodologies, I aim to sidestep as much as possible the pitfalls Naber mentions. I focus on empire to expose it, while keeping in the forefront the fact that people are not passive victims of these impositions of hegemonic power. Their actions, moreover, provide insights into empire's operations and potential lessons for further disruption.

Chapter 1 describes how the post-9/11 bioterror imaginary formed a distinct site of anti-Arab and anti-Muslim discourse. Arabs and Muslims, whom the war on terror had already racialized as culturally regressive and pathologically violent, became—through the discourse of the bioterrorist—additionally suspect, painted as lacking reason, objectivity, and morality, all the hallmarks of civilization in Western discourse. The chapter focuses on three specific figures of this imaginary: the violent and technologically backward Arab/Muslim primitive; the unstoppable Other as embodied germ; and the Third World female scientist as a mixed modern.

The bioterror imaginary fueled bioweapons proliferation, embroiling scientists, lab workers, and even surrounding communities in a dangerous research industry that exposed them to infection from lethal pathogens. The anthrax mailings represented as much the pattern of white male violence as the revelation that lab mishaps are far more commonplace a problem than intentional deployment of bioweapons. Chapter 2 focuses on the harms generated by the U.S. biodefense industry, from accidental exposures to U.S. weapons buildup; it discusses the way that the raced and gendered bioterror imaginary—in concert with discourses of technoscientific progress—serves to mask and minimize biodefense's consequences.

Bioterrorism preparedness efforts garnered some strange bedfellows. Public health's purview over disease control led to the domain's involvement in national security—not only as sentinels for germ attacks, but as partners in more overtly militaristic endeavors such as preemptive vaccination against smallpox in preparation for U.S. invasion of Iraq. This enlistment was, in the view of many health personnel, a perversion of public health's mandate of caregiving; it also created unnecessary health risks to vaccine recipients in the form of the vaccine's adverse events (side effects). Chapter 3 discusses the role the U.S. national security apparatus carved out for health workers and health care infrastructure, as well as the mobilization of public health's discursive resources. The latter includes the legitimacy bioterrorism preparedness garners when couched as in part a health care endeavor (instead of solely a military one). It also includes U.S. government and corporate media deployment of feminized imagery to conjure U.S. vulnerability: images of U.S. white women as beneficiaries of U.S. bioterror preparedness measures.

The war's tentacular reach into public health provided the entry point for greater U.S. influence over "global health security," namely, the control of disease spread worldwide—whether the result of intentional biological attacks or naturally arising outbreaks. Chapter 4 discusses the move from a focus on bioterrorism preparedness to pandemic preparedness after the emergence of H5N1 flu in Asia at the end of 2003. H5N1 became the center of a pandemic imaginary that, like the bioterror imaginary, was rooted in racial, Orientalist tropes about Asia—as uncivilized disease receptacles. This imaginary both reflected and bolstered U.S. dominance over global health infrastructure. China and Indonesia, two countries heavily afflicted with H5N1 flu, would challenge the global control of the United States and its elite allies over the pandemic flu infrastructure (surveillance, research resources, and access to treatments). Both countries reached a degree of success, illustrating a key site for understanding methods to unsettle and disrupt U.S. bio-imperialism.

1

The Making of the Technoscientific Other

Tales of Terrorism, Development, and Third World Morality

Bioterrorism is a real threat to our country. It's a threat to every nation that loves freedom. Terrorist groups seek biological weapons; we know some rogue states already have them....It's important that we confront these real threats to our country and prepare for future emergencies.
—President George W. Bush, "President Signs Public Health Security and Bioterrorism Bill," June 12, 2002

The bioterrorist is an active agent, a sophisticated hybrid of primitive and modern who seizes "our" biotechnology—a symbol of American modernity and economic might—and transforms it into a political weapon.
—Historian of medicine Nicholas B. King, "The Influence of Anxiety" (2003, 438)

Biological warfare invokes long-standing anxieties about disease spread; the use of infectious diseases such as anthrax, smallpox, and plague as "biological weapons" conjures a picture of widespread devastation. Such a weapon, comprising living organisms, can spread indiscriminately and unpredictably, and thus may produce prolonged and untold suffering. In fact, the morality of germ weapons has drawn debate since their earliest usage. The development of large-scale national programs during both world wars was followed by contentious dispute among the military, politicians, and scientists. Science writer Robin Clarke (1968) tracked the range of views: some highlighted that biological warfare entails a low proportion of casualties and disability and is thus more humane than other methods of waging warfare, while others argued that it disproportionately harms civilians and the weakest of the target population and that germs should only be approached in medical terms—that is, as something to be eliminated (not deployed for warfare).

Opposition to biological weapons ramped up in the post–World War II period, when scientists internationally began to oppose all three forms of unconventional weapons as inhumane—nuclear (because of their use against Japan), chemical (used against Vietnam), and biological, by virtue of their grouping with chemical weapons. By the mid-1960s, the UK, among other nations, sought to expand the targets of existing prohibitions. The 1925 Geneva Protocol had banned first use of chemical and bacteriological—but not microbiological and other biological—weapons or their toxins. These were included in a new, stricter international ban on offensive biological weapons programs—the 1972 Biological Weapons Convention (BWC). Its full moniker was the Convention on the Prohibition of the Development, Production and Stockpiling of Bacteriological (Biological) and Toxin Weapons and on Their Destruction. The text of the ban described biological weapons as "repugnant to the conscience of mankind"—cinching the view of biological warfare as immoral. In the United States, President Richard Nixon had instituted a ban a few years prior, in 1969, a result of concerns among scientists, politicians, and military commanders about not only the ethics of biological weapons, but also their questionable efficacy in comparison with the proven power of nuclear weapons (Cole 1997; Tucker and Mahan 2009).[1]

The dwindling of large-scale biological warfare programs was followed by the implementation of defensive programs. In the United States, biodefense ebbed and flowed under different administrations, getting a substantial boost under Clinton. In the late 1990s Clinton instituted far-reaching domestic preparedness measures and monitored carefully the weapons capacity of the former Soviet Union and other nations and groups. During this renewed attention, Clinton rallied moral discourse—particularly potent in the wake of the post-sixties ban—to code nations such as Iraq (who were accused of possessing offensive biological weapons) as uncivilized: Clinton described the UN, which was engaging in weapons inspections of Iraq, as "the eyes and ears of the civilized world" (Clinton 1998). This framing of biological warfare in moral terms would continue in the Bush era.

In a speech to the United Nations elaborating the rationale for the war on terror, President Bush stated: "Terrorists are searching for weapons of mass destruction. . . . They can be expected to use chemical, biological and nuclear weapons the moment they are capable of doing so" (Bush 2001d). This construction of certain groups as prone to violence, as if they lacked moral restraint, was central to how "terrorism" became the primary marker of Arabs, Muslims, and the Middle East in the post-9/11 period. National security narratives positioned Arabs and Muslims as the primary antagonists, and discourses about biological warfare played a pivotal role in shaping anti-Arab and anti-Muslim sentiment. Before elaborating on this biological warfare component of U.S. narratives on Arabs, Muslims, and terrorism, I turn to the origins of this sentiment in the late Cold War period.

Orientalism, Gender, and the War on Terror

In 1953, the U.S. CIA backed a coup to topple Iran's Prime Minister Mohammed Mossadegh, who had taken control of Iran's ample oil reserves. This was one of numerous instances where the U.S. government interfered in the governments of Middle Eastern nations for the sake of U.S. interest in the region's oil in the period after World War II. U.S. interventions in the Middle East intensified during the 1967 Arab-Israeli War when the United States definitively backed Israel as its strategic ally in the region. As the United States and Europe engaged in economic and military interventions for oil resources in the Middle East, Arabs and Arab Americans protested U.S. imperialism. To suppress this resistance, the FBI spied on Arab Americans via Operation Boulder in 1972 (Naber 2000). At the same time, popular media such as TV narratives depicted Arab men as greedy oil sheiks and bearded terrorists, and Arab women in harems and as belly dancers (Shaheen 1984). These narratives served to portray Arab peoples—and the Middle East—as backward, uncivilized, exotic, and dangerous.

As is well described by Edward Said (1978), depictions of the "Orient" have served as ideological tools aiding empires since the late eighteenth century—first the British and French, subsequently the United States. The role of knowledge production in the colonial project has relied primarily on producing images of the "Orient" in dualistic terms that serve to affirm Western[2] cultural superiority—for example, backward/civilized, superstitious/scientific, irrational/rational, archaic/modern, evil/good, violent/peaceful. Said called this Orientalism, tracing its production to the discipline of geography, biblical texts, armies, colonial administration, and scholarship.

In the United States, Orientalism linked up with racism to not only bolster U.S. policy in the Middle East, but also, as Arab American studies scholars have demonstrated, to negatively racialize Arabs within U.S. society. Arabs, who had been immigrating to the United States since the late nineteenth century, experienced a downgrading of social status during the post–World War II period via public policies, mainstream representations, social patterns of discrimination, separation, and exclusion (Cainkar 2008). By the 1960s, Arab Americans had lost many of the privileges that their previous near-white status had afforded them,[3] and were increasingly viewed as innately culturally different from—and inferior to—whites. Cultural essentialism, then, was fundamental to this racialization of Arabs. It marked a distinct mode of racialization that drew on existing Orientalist dualisms, but contrasted with the biologically, phenotypically based racism that has marked blacks and other people of color since the advent of scientific racism in the late 1800s.[4]

No cultural marker has been as key to Arab racialization as Islam. The U.S. state has viewed Islam as a symbol of political subversion since the 1930s, when black Americans began to turn to it as a tool of black liberation. U.S. state surveillance and prosecution of black Muslim organizations escalated during the 1960s, when the FBI utilized aggressive counterintelligence, most notably COINTELPRO, to spy on and infiltrate the Nation of Islam and other black liberation groups that threatened the white supremacist status quo (Curtis 2013). During the late Cold War period, the United States began to focus on Muslims outside the country—the Islamic Republic that took power in Iran in 1979 and the increasingly powerful Islamists in the 1990s, both of which vociferously criticized the United States and its global hegemony.[5] The United States began to view Islam as a threat and also as the main signifier of Arabs and Arab Americans (Hatem 2011). Nadine Naber (2000) describes how popular films such as *Not without My Daughter* (1991) and *The Siege* (1998) portrayed Islam as the driver of Arab backwardness and violence. Although there had been some conflation of "Arab" and "Muslim" (both in negative terms) in earlier colonial contexts,[6] the late Cold War period saw the solidification of the Arab/Muslim terrorist figure.

The late Cold War construction of an Arab/Muslim threat intensified cultural essentialisms that pivoted on the notion of a clash of values, ideology, and religion (Jamal 2008; Muscati 2002; Shyrock 2008). This ideology continued to gain clout throughout the post–Cold War period. Conservative political scientist Samuel Huntington's treatise in 1993 on the "clash of civilizations" became wildly popular. It posited that Arabs and Muslims were the cultural Other of the West—inherently and incommensurably culturally different. This cultural essentialism racialized Arabs and Muslims, but in a seemingly race-neutral manner. As Arab American studies scholar Louise Cainkar (2008) has argued, the focus on essentialized cultural and religious differences effectively obscured the racial dimensions of this worldview in a post–civil rights era where blatant racism was no longer acceptable (48).

This thesis of inalterable cultural difference would culminate in the war on terror: rather than pursue criminal prosecution of the September 11 perpetrators, the Bush administration used the attacks as an opportunity to augment U.S. presence in the Middle East. On October 7, 2001, the United States launched an invasion of Afghanistan. This was followed by an invasion of Iraq in 2003. Domestic counterterrorism policies worked in tandem with this military action to target Arabs and Muslims: on October 26, 2001, President Bush signed into law the USA PATRIOT Act (Uniting and Strengthening America by Providing Appropriate Tools Required to Intercept and Obstruct Terrorism Act), which enhanced the ability of the U.S. government to surveil and criminalize foreign nationals, disproportionately impacting Arabs and Muslims.[7]

Suad Joseph, Benjamin D'Harlingue, and Alvin Ka Hin Wong (2008) demonstrate how Arab Americans and Muslim Americans were represented in the mass media after the September 11 attacks—as more intimately tied to their countries of origin than other immigrants are, and as more tied to their countries of origin than they are to the United States (234). They cite *New York Times* journalist Laurie Goodstein's September 12, 2001, article "In U.S., Echoes of Rift of Muslims and Jews" as an example: Goodstein depicts Muslims in the United States as affected by terrorist attacks because they have "kin in the Middle East" and thus have "struggled to assert their identities as loyal Americans" (241). Such mass media portrayals intensified a mainstream view of Arabs and Muslims as dangerous and unassimilable, and as having foreign—and thus assumed potentially terrorist—ties. In choosing to perpetuate notions of Arabs and Muslims as an enemy within, these media portrayals largely sidestepped analysis of U.S. imperialism and the social and historical context of violence against the United States by neofundamentalist Islamic groups like Al Qaeda.

U.S. government and media marking of Arabs and Muslims as terrorist Others fueled societal violence. Hussein Ibish, senior resident scholar at the Arab Gulf States Institute in Washington, has documented an increase in hate crimes, violent incidents, and discrimination in airline passenger seating, employment, housing, and so on in the year following the September 11 attacks (2003). Meanwhile, the Bush administration, despite waging war against Arabs and Muslims abroad and targeting them within the United States, issued disingenuous statements against the violence Muslims faced from fellow U.S. residents. As early as September 17, President Bush had urged that Muslims "need to be treated with respect" and "must not be intimidated in America" (Bush 2001b). This denouncement, in addition to being rather mild, also reproduced the notion of a binary between the United States and Muslims: the statement started with Bush addressing "both Americans *and* Muslim friends and citizens" as if the two were nonoverlapping categories. His statement also linked Islam with terrorism: "The face of terror is not the true face of Islam" (Bush 2001b; emphasis added). In making the point that not all Muslims were terrorists, Bush indelibly linked Islam with terror—until proven otherwise.

The clash of civilizations worldview emerged more forcefully in the speeches to follow. President Bush stated in November 2001: "This new enemy seeks to destroy our freedom and impose its views. We value life; the terrorists ruthlessly destroy it" (Bush 2001c). Here, a clear demarcation is being drawn between the "civilized" and the "barbaric." The notion that death and violence are values, and not tactics, reinforced the notion of a civilizational binary dividing terrorists (read: Arabs and Muslims) from nonterrorists (read: the United States and the West in general). Puar and Rai (2002) describe how, in arenas as diverse as academic discourses of "terrorism studies" to popular TV shows, there was a generalized post-9/11 sentiment about a "terrorist psyche" and "terrorist culture" that purportedly reflected a fundamental divergence from U.S. norms of morality.

News media further entrenched this clash of civilizations worldview. In 2003, a journalist from the *Wall Street Journal* voiced his support for the war on terror: "If Mr. Bush had not declared war on global terrorism and had not declared his willingness to strike first," then "the civilized world would be staring down the *test tubes of barbarism*, with no better strategy than waiting for some Saddam [Hussein], Kim Jong Il, Osama bin Laden or any of the other nihilists along the spectrum of WMD acquisition to annihilate large numbers of some nation's civilian population" (Henninger 2003; emphasis added). The phrase "test tubes of barbarism" served not only to evoke alarming germ-inflected imagery but also to reveal the dualistic terms by which the West/Global North justifies its waging of imperialist wars through constructions of itself as civilized, in contrast to those it characterizes in opposite terms (typically those designated by the overlapping categories of the "East," Global South, and formerly and currently colonized countries). In April 2004 the *Los Angeles Times* featured the following commentary: "Muslim leaders need to accept the fact that their religion has been infected by a virus that embraces death and delivers it with cold, unfeeling calculation" (*Los Angeles Times* 2004). This quote mirrored Bush's rhetoric: it disarmingly gestured to an Islam that may have been innocent at one time, while employing the metaphor of uncontrollable infection to depict Islam, and thus Muslims, as death-loving and nefariously calculating.

In addition to being markedly racialized and Orientalist, the war on terror was distinctly gendered. Long-standing tropes of Arab and Muslim masculinity as atavistic, misogynist, hypersexual, and ultimately dangerous found renewed vigor in post-9/11 vernacular as well as academic discourses—the Arab/Muslim terrorist was primarily articulated as a masculine threat.[8] This was reflected in the early post-9/11 actions of U.S. law enforcement agencies. The Department of Justice, backed by the PATRIOT Act, instituted the secret detention of thousands of mostly Arab and/or Muslim men based on minor citizenship status violations, while other foreign nationals in similar situations were not picked up.[9] This was followed one year later, in September 2002, by the National Security Entry-Exit Registration System ("Special Registration"), wherein the Immigration and Naturalization Service (INS) instituted point-of-entry registration of foreign nationals from five Middle Eastern countries (Iran, Iraq, Syria, Libya, and Sudan)[10] as well as the registration of noncitizen males over age sixteen whose national origin was from twenty-five countries, all but one of which were Arab and/or Muslim-majority nations.[11] This resulted in the detention and deportation proceedings of over 13,000 (there were over 80,000 registered), even though none were charged with terrorism, and fewer than 200 with any criminal activity whatsoever (Ibish 2003; Vanzi 2004).

Representations of the Arab/Muslim woman have constructed her in opposite terms to the misogynistic, violent Arab/Muslim man—she is passive, oppressed by the men in her culture, and in need of "saving" by the West (Moallem 2002; Nayak 2006).[12] In spite—or because—of this portrayal, Arab and Muslim women too have been subject to violence and discrimination by U.S. institutions (airline profiling, employment discrimination, public violence, verbal and physical harassment), as well as by the larger public (assaults such as having their hijabs pulled off) (Ibish 2003). Naber (2008) notes: "A general consensus among community leaders was that federal government policies disproportionately targeted men while hate crimes and incidents of harassment in the public sphere disproportionately targeted women" (293).

The Violent Primitive

The violent Arab/Muslim male terrorist trope appeared in myriad places during the war on terror—from the statements of national security officials, pundits, and legislators, to news media and popular culture, as well as in academic literature.[13] The prolific cartoonist Michael Ramirez, a regular contributor to neoconservative publications such as the *Weekly Standard*, illustrated this trope in

FIG. 1. A cartoonist's sketch of the perpetrator of the 2001 anthrax mailings. From *Los Angeles Times*, October 13, 2001, B21. (By permission of Michael Ramirez and Creators Syndicate, Inc.)

a *Los Angeles Times* drawing of a bioterrorist in late 2001, just after the anthrax mailings (see figure 1).

The marking of this figure as an Arab / Muslim male terrorist is evidenced by the caricatured features (the shape of the nose and head, the headdress), the setting (the cave-like dwelling), and the caption referencing a "warrior of the jihad." As Naber (2008) describes, Arabs and Muslims became marked by a wide range of signifiers: name (e.g., Mohammed), dark skin, dress (e.g., a head-scarf or a beard), and nation of origin (e.g., Iraq or Pakistan). This is indicative of post-9 / 11 Arab / Muslim racialization that centers on a combination of physical attributes, religious identity, and (foreign) nationality, but also the fact that "Arab/Muslim" had become an expanded racialized category revolving around being "terrorist-looking" or, synonymously, "Muslim-looking." This racialization lumped together several incongruous subcategories of racialized Others:[14] Arabs and Iranians, including Christians, Jews, and Muslims; Muslims from Muslim-majority countries; and persons who are perceived to be Arab, Middle Eastern, or Muslim, such as South Asians, including Sikhs and Hindus (Naber 2008).[15] Ramirez's visual intertwines these themes to sensational—and racial—effect.

Moreover, Ramirez invoked the gendered dimensions of this racialization—specifically the Arab/Muslim male as misogynist. Queer theorist Jasbir Puar and literary theorist Amit Rai (2002), among others, have outlined how the misogyny attributed to Arab and Muslim men (i.e., they are typically portrayed as oppressing their female counterparts) serves to pathologize Arab and Muslim cultures as violent and backward (and thereby rationalize the brutal targeting of Arabs and Muslims through the war on terror).[16] The cartoon image pathologized Arab and Muslim masculinity by highlighting not only misogyny, but also cowardice—the figure's targeting of women and children (the wording on the mailbox) as well as the sarcastically phrased caption, "The great and oh so brave warrior of the jihad . . ." Ramirez's depiction of women and children as targets was particularly misleading considering that the anthrax mailings of 2001 were not actually sent to women and children, but to politicians (all of whom were male) and news media venues. Nevertheless, the inclusion of women and children as a trope was purposeful: it suggested a cowardly perpetrator who reprehensibly—and immorally—targets the nation's most weak and vulnerable (i.e., "women and children"). The cartoon, then, set itself up as unmasking a "warrior of the jihad" to be a coward motivated by personality and, moreover, cultural flaws.[17] This invoking of cowardice in concert with misogyny fortified the notion of a deviant Arab/Muslim male symbolizing an uncivilized and violent culture.

Ramirez's reproduction of the dominant narrative that Arabs and Muslims are backward and violent in effect undermined any reading of Arabs and Muslims as casualties of U.S. imperialism, the latter obscured from view. His cartoon echoed larger discourses that constructed Arabs and Muslims as culturally violent, rather than as rational actors who at times use violence as a political response to hegemonic power—from the liberation struggles of black American Muslims in the 1960s to the Islamist repudiation of U.S. influence in the Middle East since the 1990s.[18] As history and politics scholar Mahmood Mamdani (2004b) aptly states: "This [mainstream Western] history stigmatizes those shut out of modernity as antimodern because they resist being shut out" (19).

SCIENCE AND REASON

The cartoon depicted the actual act of bioterrorism in exceedingly simple terms—the dropping of a licked, sealed envelope in the mail. This impression of an easy delivery mechanism was further conveyed by the cave-like dwelling—a small, dingy apartment with sheets for curtains connoted simplicity as well as primitivity. The omission of both scientific expertise and a scientific setting from the picture belied the processing and cultivation entailed in weaponizing germs—which typically require high-security, well-equipped labs, and in the specific case of the anthrax mailings was in fact traced to a U.S. laboratory.

As mentioned, the West has mobilized Orientalist logics since the colonial period to mark itself as possessing rationality, scientific thinking, and advanced technology, and as superior because of these attributes. In turn, the West marks the East as irrational and backward, and as constitutionally unable to produce the tools and products of an advanced civilization—that is, science/technology (as well as other features of Western cultures such as industrialization and capitalist development). The West has used this narrative to couch colonial exploitation as a charitable attempt to help inferior cultures develop and progress toward civilizational advancement (Adas 1989, 205, 220; Macleod 1993, 123). As Stuart Hall (1992) describes: "The West was the model, the prototype and the measure of social progress. . . . And yet, all this depended on the discursive figures of the 'noble vs ignoble savage,' and of 'rude and refined nations' which had been formulated in the discourse of 'the West and the Rest.' . . . Without the Rest (or its own internal 'others'), the West would not have been able to recognize and represent itself as the summit of human history" (221).

Thus, Western historians of science have constructed a genealogy of science in a linear fashion from within Europe that excludes not only traditions emanating from other locales, but also the incorporation of non-European sources into modern science in Europe. Specifically, the West has ignored and dismissed rich histories of science originating from Arab and Islamic thinkers. Mamdani (2004b) has described how Western historians have dismissed Arabic-writing scientists during the classical age of Islam (from the eighth to the thirteenth century) and constructed them as merely preserving classical Greek science and passing it on— without any significant contribution—to Renaissance Europe (from the fourteenth to the seventeenth century).

The Ramirez cartoon reflected this Orientalist view: its picture of mundane primitivity (in the terrorist's ease of waging biological warfare) combines features of Western narratives of science and progress with the savagery attributed to the Arab/Muslim male that I have already outlined. Thus, Ramirez's caricatured Arab/ Muslim man lacks scientific expertise and its associated trappings— reason, rationality, and objectivity. Such a figure feeds into anti-Arab/anti-Muslim discourse, insinuating that Arabs and Muslims are of lesser intellect and scientific capacity, producing yet another terrain of their dehumanization.

The Unstoppable Germ

A second type of figuring of the bioterrorist emerged in U.S. national security debates on bioterrorism response scenarios. Pundits from national security and public health officials to bioethicists and journalists contemplated a new, entirely imagined, threat: the "suicide infector." Also called "suicide disease carrier," "suicide disease bomber," and "smallpox martyr," among other names, the figure denotes someone who infects themselves with disease, sneaks into the country (the United States) before showing symptoms of the disease, and consequently starts an epidemic.[19] The self-infected terrorist became a subject of U.S. security discussions on possible Iraqi biological weapons—namely, smallpox—possession from late 2001 to early 2003. In these discussions, the suicide infector signified the degree of potential Iraqi threat, helping bolster the U.S.

rationale for preemptive action—both the invasion of Iraq and the vaccination of members of the military and public health sector against smallpox.[20]

The hypothetical nature of the figure made it possible for pundits to experiment with ideas about bioterrorism and bioterrorists. A RAND Corporation report described a study conducted by its Center for Domestic and International Health Security that focused on "suicide attackers who ride mass transit spreading the virus" as one of several "feasible smallpox attack scenarios" (2003). A *New York Times* journalist described the possibility that "a smallpox epidemic could begin with a single infected person—a 'smallpox martyr,' in the terminology of bioterrorism experts—simply walking through a crowd" (Stolberg 2001). Such scenarios made a suicide infector attack appear probable and, moreover, as something that could occur in the most mundane of public settings.

The suicide infector derives its genealogy from the suicide bomber,[21] drawing on the latter's potency in demarcating the Orientalist binary. Sociologist of race and terror Gargi Bhattacharyya (2008) describes suicide bombing as representing, in Western discourse, "an indication of the absolute difference between 'us' and 'them,' and the instance of the boundary is seen to stem from the dysfunctional subject formation of so-called enemies of the West" (54). The suicide bomber is seen as Other because of a willingness to not only create mass, indiscriminate destruction, but also perform a monstrous act of human sacrifice. The suicide bomber's mode of warfare is, thus, designated illegitimate and uncivilized—it is not the *willingness* to give one's body over to a military cause, but the *nature* of the sacrifice that marks the terrorist as Other (as opposed to, say, the exaltation that meets the patriotic soldier). In this way, Western discourses on suicide bombing have racialized Arab and Muslim cultures as engaging in warfare that lacks a proper sense of morality and outlook on death (Amireh 2011; Asad 2007; Brunner 2007; C. Lee 2009).

The suicide infector figure builds on this notion of Arab and Muslim cultures as backward, violent, and morally questionable. To weaponize one's body with germs—thereby weathering grave illness—epitomizes the type of monstrous self-sacrifice that marks Arab and Muslim cultures as depraved in Orientalist discourse. The germ dimension, moreover, imbues the suicide infector with an element of metaphor: the suicide infector mirrors the ability of germs to contaminate everyday spaces.

The figure made its way through the mass media, often through vivid description. It even appeared in relatively obscure newspapers such as the *Naperville Sun*;[22] in late November 2001, the paper published an article covering bioterrorism response plans being explored in local health departments and other agencies, opening with a worrying scenario: "Bioterrorists in New York City, Washington, D.C, and Chicago have deliberately infected themselves with smallpox virus and are *contaminating* public, highly visible places, like subways, government buildings and shopping malls" (Pazola 2001; emphasis added). As in the previous examples, the theme of contamination invited a slippage between germ carrier and germ, but further invited the reader to dwell upon a variety of horror scenes.

The characterization of the suicide infector as contaminant built on wider post-9/11 discourse likening terrorism to a vicious disease, and terrorists to despicable vermin (C. Lee 2009; Sarasin 2006). An August 2004 article in the *Wall Street Journal* described terrorism as spreading "like a virus." Historian Philipp Sarasin (2006) notes the many explicit references to the September 11 World Trade Center attackers as uncivilized vermin by a diverse array of sources from journalists to survivors. This rhetoric echoes long-standing narratives of the Other that analogize them as germs, vermin, and parasites infecting the national body (Steuter and Wills 2009). Decolonial theorist Linda Tuhiwai Smith, in discussing the subjugation of indigenous peoples, highlights how such rhetoric serves to dehumanize—and this dehumanization has been pivotal to subjugation attempts: "To consider indigenous peoples as not fully human, or not human at all, enabled distance to be maintained and justified various policies of either extermination or domestication. Some indigenous peoples ('not human'), were hunted and killed like vermin, others ('partially human'), were rounded up and put in reserves like creatures to be broken in, branded and put to work" (Smith 2002, 26). The germ contaminant dimension of the suicide infector trades on this discourse of "humanity," suggesting an enemy so vile and wholly destructive that only the most extreme measures will be able to stamp it out.

But the suicide infector scenario mobilizes the germ analogy to further dire effect: it turns the metaphor of terrorism as contagious disease into a material reality. The suicide infector is not *like* a contaminant, he[23] *is* the contaminant. Like the suicide bomber, the suicide infector connotes an enemy who can sneak into highly public spaces without detection to perpetrate deadly acts. But the actual body of the suicide infector symbolizes more than just a dangerous, insidious presence—he is literally a bundle of living, regenerating germs that can proliferate indefinitely. The infectivity of the suicide infector, I suggest, hooks onto the notion of infiltration in post-9/11 discourses of Arab Americans and Muslim Americans as the enemy within (i.e., as loyal to their fundamentally suspect, culturally different countries of origin). The Arab/Muslim Other as germ body, in this view, represents both a cultural and a biological threat.

One *New York Times* article painted a grim picture of the figure's virulent potency:

Even before boarding his plane, the "human missile" *crisscrosses* the airport, stands in line at check-in, at the Starbucks stand, in the bathroom, at security. Whenever he coughs, some people close to him will breathe the virus in, and it will lodge in their lips and noses, and they will carry it inside them onto their own planes, passing it to the passengers directly around them. In the airport alone, experts estimate, a smallpox martyr can infect between 3 to 20 other people. And in a confined space with internally circulating air, that number could be even greater. Americans wouldn't hear anything for another two weeks as the virus incubates. Then in different corners of America, wherever those planes landed, hundreds if not thousands will come down with the "flu." Their backs will ache. Their fevers will spike.

Their skin will darken until it looks charred, and then things will really get bad. There is no treatment. By this point, a vaccine is useless. (Landesman 2002; emphasis added)

The language chosen to characterize the figure's "crisscrossing" movement invokes the flexible, arbitrary motion that germs possess, and the language of a "missile," a military weapon, linking the two to bring forth rampant, prolonged suffering. This effectively tethers the mobility of the germ to the destructiveness of a military force, inviting the image of an Arab/Muslim body bent on widespread ruin. It is a version of the primitive bioterrorist figure I described in the previous section—lacking advanced technology and expertise, but here availing himself of the most primitive of technologies—his weaponized body. In this picture, the Arab/Muslim body is quite literally reduced to a subhuman entity—savage and without capacity for reason. This is a highly fear-inducing view that links up with discourse constructing Arab and Muslim cultures as depraved and backward—there is no better symbol of depravity than a figure who chooses to cede control of his body to the primordial, chaotic, and wholly destructive will of the germ.

Most of those engaged in the debates in government and media spheres over various suicide infector scenarios and their plausibility ultimately dismissed the likelihood that a suicide infector would succeed in sowing disease. Yet, the debates, in fostering the metaphor of Arabs and Muslims as germs that spread, contaminate, and infect, served to reinforce the specter of Arab/Muslim terrorism. The image of an Arab/Muslim germ body expanded the terror imaginary into new territory and, I suggest, aided a post-9/11 U.S. security apparatus invested in fomenting racial terror—for the purpose of validating extreme actions in the name of defense, namely, unprovoked invasion of Iraq.

The Semi-Modern Scientist-Terrorist

The final figuring I dissect departs from the trope of the a-scientific, primitive Other; in contrast, this third figure represents a more complex notion—the semi-modern bioterrorist. This educated, often Western-trained Arab / Muslim scientist has attained some of the material attributes of Western civilization—that is, scientific and technological expertise—but without the requisite cultural characteristics (i.e., norms, values, morality) that purportedly define Western civilizational superiority. This trope reached fruition, like the suicide infector, in relation to Iraq, but drew more on the reality of the country's scientific and technological industries.

Drs. Rihab Rashid Taha al-Azawi and Huda Salih Mahdi Ammash were two high-level Iraqi scientists. Dr. Taha earned her PhD in microbiology in the mid-1980s from England's East Anglia University, returning to Iraq afterward to work at the Iraqi chemical weapons plant al Muthanna. Dr. Ammash earned her PhD in microbiology from the University of Missouri in 1983, and upon returning to Iraq eventually became dean of the College of Education for Women and dean of the College of Science at Baghdad University, head of Iraq's microbiology society, and a member of the Baath Party's command council in May 2001. She had researched and published widely on the effects of the U.S. government's bombing of Iraq during the First Gulf War, namely, the carcinogenic effects of depleted uranium.

Both were arrested and held in detention for two and a half years by U.S. occupation forces in Iraq, from May 2003 to December 2005 (coincident with the Iraq War), for their suspected involvement in Saddam Hussein's biowarfare program—although neither was officially charged. Iraq's biological weapons program, active during the Iran-Iraq War from 1980 to 1988, had been destroyed in the mid-1990s after inspections following the first U.S. intrusion into Iraq (the First Gulf War in 1990). Iraq was again inspected in 2002, with the UN confirming the lack of any active operations.[24]

The U.S. government was not forthcoming with details about their claims of the scientists' involvement in Saddam Hussein's biowarfare program. Much of the publicly accessible information about them resided in mainstream news reports (and the occasional military report that surfaced), which frequently painted them in guilty terms. A *USA Today* article described Taha as "the best-known of Iraq's biological weapons scientists" who "headed Saddam Hussein's government lab that weaponized anthrax in the 1990s" (Leinwand and Parker 2003). A *Chicago Tribune* article stated: "U.S. intelligence places her [Ammash] at the heart of Iraq's efforts to develop biological weapons" and as someone "who has been described as one of the pillars of the country's [Iraq's] weapons program" (Swanson 2003). Many progressive news outlets, on the other hand, questioned the veracity of the claims connecting these scientists to biological weapons, especially after sources from within the government as well as independent sources debunked U.S. allegations of Iraqi WMD possession (Scheer 2005; Spinoza 2003). Advocates challenged the detainment of Dr. Ammash in particular, suggesting it was U.S. retaliation for her published studies criticizing the effects of the U.S. government's bombing of Iraq. U.S. groups such as the prestigious American Association for the Advancement of Science (AAAS) and the American Association of University Professors called for her release (AAAS 2005; Hunter and Salama 2006).

In the end, both were released, without charges ever having been brought against them. The actual culpability of these two scientists in relation to biological warfare is difficult to sort out, [25] but what is clear is that their detention was part of an unjustified U.S. project, namely, the invasion of Iraq. Like other Iraqis, they were sacrificed to the goals of U.S. empire. Their detention caused them undue suffering: Ammash had a recurrence of breast cancer while in detention (Jaschik 2005; "Iraq's Jailed" 2005). She passed away ten years later, in November 2016; she was in her early sixties. Their detention, moreover, produced harms on a discursive level affecting perceptions of Iraq as well as Arabs and Muslims more generally—they became the face of an Iraqi bioweapons threat and also the narrative of an evil, lawless, and duplicitous regime led by the quintessential Arab/Muslim villain, Saddam Hussein.[26]

News coverage often highlighted that Taha and Ammash were UK- and U.S.-educated, respectively, for their graduate degrees.

One representative text, for example, an *NBC News* biographical piece about Dr. Taha, contained a section titled "Copying the U.S." stating that "Taha's *western experience* was more than helpful to its [Iraq's superweapons programs'] success. For example, Taha knew that Iraq could order anthrax from specimen houses in the west, including the United States" (Windrem 2004; emphasis added). First, the word choice "copying" invoked ideologies of U.S./Western knowledge and culture as superior and desirable to Iraqis seeking to advance their culture. Second, the emphasis on Taha's Western experience signaled another trenchant narrative: the Third World scientist following the path of Western advancement.

When colonized countries shrugged off the yoke of colonialism after World War II, they found themselves ensnared in neocolonial arrangements that perpetuated their impoverishment and underdevelopment. The Cold War realignment of the world system dominated by the United States and its First World allies on the one hand, and the former Soviet Union and its allies on the other, produced the Third World, that is, countries caught in the crosshairs of the battle between the two camps. Concomitantly, colonial narratives of the inferiority of colonized people, their cultures, and their knowledge systems (i.e., science) evolved into Cold War narratives of modernization and development that presumed the racial and/or cultural deficiency of the Third World. These narratives promoted the ideology that the plight of the Third World was the product of internal deficiencies rather than ongoing colonial exploitation. The solution laid out in these narratives was for Third World scientists to follow a Western path reliant on advanced industrial technologies to prosper (Philip 2015; Bhuiyan 2008; Escobar 1995).

Even though the article invoked the notion of "western experience" as the path to Third World advancement, careful attention to what the article denoted by "western experience" in the sentence that followed—Taha knew where the West housed its anthrax—reveals that "western experience" is quite overstated. Such a gloss reflects a deeply entrenched subtext—the assumption of Western superiority. It is a notion so taken for granted that the article need not even explicitly reference it—a vague gesture to "western experience" suffices.

The quote also intimates that Iraqi acquisition of "western experience" poses a threat—it is linked to Iraqi superweapons. The end of the Cold War and the advent of U.S.-led global capitalism had produced a severe power asymmetry wherein the United States turned to subduing what were significantly smaller threats globally. Yet, U.S. national circles became concerned with these smaller threats in the backdrop of the vast interconnectedness and mobility wrought by global capitalism, which had made both weaponry and advanced technologies much more accessible to even the most disempowered nations and groups (Cecire 2009). Iraqi scientists like Taha embodied this threat to U.S. hegemony—she had accessed Western technology and could now use it against the United States. One Department of Defense–affiliated study on Taha from 1999 (Taha had been the subject of UK and U.S. surveillance since the mid-1990s) quoted Andrew Koch, an analyst at the Center for Defense Information in Washington: "It is the scientists who are the key to this [weapons capacity]. . . . As long as Iraq maintains the *brainpower* to do this . . . over the long term you can't stop them" (Brian Anderson 1999, 21; emphasis added). As with the news quote above, connections with the West were depicted as a source of intellectual (and technological) power, leading to the image of Western-educated Iraqis who were both advanced and potentially dangerous.

In suggesting that "western experience" and technologies can be dangerous in the hands of the non-Western Other, these quotes intimate an ambivalence toward educating and equipping Iraqis. This logic reflected the underlying reality of U.S. empire—that the United States had less of a problem with Iraqi possession of weapons of mass destruction (as the United States had after all helped develop and furnish these weapons to Iraq while it was an ally),[27] and more of a problem with Iraqi ability to use them without direction from the United States or, worse, against the United States.

Returning to the 2004 *NBC News* biographical piece on Dr. Taha, a later passage used the trope of the Third World scientist as threat to mark Iraq as a semi-modern culture lacking the civilizational attributes to handle advanced technology: "Dr. Rihab Rashida Taha would rank among the most important of a new breed of Third World weapons designers—highly nationalistic, western-educated and willing to violate any international norms or scientific ethics" (Windrem 2004). This quote juxtaposed Western-derived knowledge with the supposed backward immorality of the Third World. In this clash of civilizations imaginary, the Third World scientist becomes a potent hybrid: the scientist-terrorist who combines scientific know-how with a pathological drive for violence.

THE "GOOD," WESTERN-EDUCATED MUSLIM WOMAN GONE WRONG

Discourses of gender heavily shaped the way Drs. Taha and Ammash and their actions were portrayed across a variety of news media and specialized sources. The press dubbed them with sensationalist epithets: Ammash as "Mrs. Anthrax" and Taha as "Dr. Germ" as well as "world's deadliest woman." Depictions focused inordinately on their physical appearance, and in comparison with each other: Taha was the dowdy one, Ammash the fashionable, elegant one (see, for example, Windrem 2004). As outlined by gender studies scholar Robin L. Riley (2004), they were at times subject to dichotomized caricatures—as either the brains and masterminds of the Iraqi bioweapons program, or as dupes and fronts for the real bosses. While the latter depiction conforms to the ubiquitous trope of the passive Arab / Muslim woman, the former confounds this image.[28] This interweaving of gendered and racialized tropes in portrayals of the two scientists warrants a closer look.

Since the 1990s, Taha had consistently been depicted in terms of emotional imbalance: that she "would explode into a rage, shouting and tossing chairs" (Brian Anderson 1999), "would turn on the tears and even throw small tantrums in sessions with U.N. investigators" (R. Wright 1995), or "would stammer and cry when confronted with uncomfortable facts" (Windrem 2004). Ammash was described not in terms of histrionics, but in animalistic terms: for example, as "a lioness ready to pounce on her prey" or as a wily "fox" (see, for example, Swanson 2003). The portrayals of Taha as hysterical and Ammash as animal-like reflect the way women of color and colonized women have been depicted as wild and out of control—in order to rationalize their subjugation.[29] Whether devious mastermind or wild savage, these various gendered portrayals of Taha and Ammash all served the same function: they marked them as dangerous. They also, I contend, invoked a new version of Arab/Muslim threat—the good Arab/Muslim woman gone wrong.

Asian American feminist studies scholar Sunaina Maira has dissected the post-9/11 imperialist narrative that dichotomizes Muslims into "good" versus "bad." Maira built on Mamdani's theorizing of the "good/bad Muslim"—that is, hegemonic U.S. discourse juxtaposing "bad Muslim" terrorists against "good Muslims" willing to join the West in fighting against terrorism—by elaborating its gendered and moral dimensions. She demonstrates the way that Muslim masculinity is constructed as a "bad" site of potential radicalism and violence, whereas Muslim women are seen as a "good" site of potential civilizing via Western intervention. In this story, women can cultivate moderation (as opposed to religious zealotry), individualism, feminism, and ultimately, alignment with the humanitarian premise of Western imperialism (Maira 2009). Westernization, then, can emancipate Muslim women from their purportedly backward, misogynist cultures. Women are, in this narrative, the only Muslims who can be civilized and made Western.

I read U.S. depictions of Drs. Taha and Ammash as invoking the trope of the wild, savage woman—but within the terms of the good Arab/Muslim woman narrative. Taha and Ammash represent the implicit denouement of this narrative—the "good," Western-educated Arab/Muslim woman who inevitably goes wrong (becoming the wild, savage woman). U.S. representations of Taha and Ammash invoked the narrative of them as potentially civilizable "good" Arab/Muslim subjects, only to set up for their eventual failure: Taha and Ammash, given all the advantages of exposure to the West, obtaining high-ranking positions in science (and government)—symbols of intellectual advancement and power—in the end reject Western values of peace and civility and use their Western education and scientific expertise for violence (i.e., bioweaponry). They can obtain advanced science and technology (assumed to be Western), but not its handmaidens of prudence and rational restraint. In this narrative even the most redeemable Arab/Muslim subject—female, Westernized, and scientific— remains obstinately uncivilized, as fixed by an immoral nature as the crazed male terrorist figure.

News depictions of Taha also seized upon feminism to further meditate on cultural progress in Iraq. The *NBC News* piece on Taha stated, "She [Taha] has also been held up as an example to Iraqi women interested in science, in spite of a career devoid of any accomplishment other than the development of germ warfare" (Windrem 2004). This quote recapitulated the notion of Arab/ Muslim inferiority through the claim that Iraqis hold up an immoral weapons scientist as a model of female progress, thereby gesturing toward Iraqis' misconstrued feminism. Interestingly, in the two mentions on Ammash in the article, there was no note of her having headed Iraq's national microbiology society or having been dean of the College of Education for Women at the University of Baghdad and dean of the College of Science; the article highlighted only her pro-Iranian activism and her high fashionability. Perhaps Ammash's non-biowarfare accomplishments would contradict the author's attempt to build an archetype of a weapons-wielding Iraqi female scientist. In effect, the author's focus on Taha, whose postgraduate career consisted mainly of her subsequent work at the Iraqi chemical weapons plant al Muthanna, made for an easier characterization of an Iraqi female gone wrong.

Such a characterization of Iraqi female progress—or lack thereof—relies on ideologies of U.S. exceptionalism that portray Iraq as trying, but failing, to be modern. Critical race and ethnic studies scholar Chan-Malik has discussed how this exceptionalist picture of the United States works—in dialectic fashion—through

Orientalist logics: "the liberal vision of a free, feminist, and multicultural [U.S.] nation as a fundamental necessary counterpart to the decidedly unfree, antifeminist, and antidemocratic ideology of Islamic Terror" (2011, 134). Returning to the line I quoted above, the news piece highlighted a common colonial metric of the cultural progress of the Other—women and feminism. The article also enacted the assumption of the United States as cultural standard—rightful judge and arbiter of the moral and cultural progress of the Other (which it marked as failing). In marking Iraqis as misunderstanding feminism (holding up Taha as a model of female progress), the news piece, I argue, invoked the notion of the semi-modern Other who grasps the need for feminism, but lacks the cultural attributes to fully succeed at attaining it and, moreover, is so backward as to mistake the feminist for the scientist-terrorist!

This Orientalist depiction of Iraq was elaborated even more fully in a piece by human rights advocate and journalist Nicholas Kristof (2002), who centered Taha in an article extolling Iraq's empowerment of women in contrast to its regional neighbors. After outlining diverse arenas of women's achievement (from bossy behavior, army representation, sports, education, and workplace equity), he highlighted Taha in particular by placing her at the forefront of the text and making her one of only two women whose names were spelled out (the other being Nadia Yasser, the captain of the Iraqi national women's soccer team): "Iraqi women routinely boss men and serve in non-combat positions in the army. Indeed, if Iraq attacks us with smallpox, we'll have a woman to thank: Dr. Rihab Rashida Taha, the head of Iraq's biological warfare program, who is also known to weapons inspectors as Dr. Germ" (Kristof 2002). In placing Taha among various examples of Iraqi female progress, and suggesting that Taha was one of Iraq's top female role models despite her involvement in biological warfare, Kristof unequivocally conveyed the notion that Iraqis cannot distinguish between progress and evil.

Kristof, moreover, used this powerfully Orientalist depiction of Iraq to paint a broader swath of Arab/Muslim Other. Earlier in the article he had described Iraq's progressivism—but only relative to its regional neighbors: "More broadly, in a region where women are treated as doormats, Iraq offers an example of how an Arab country can adhere to Islam and yet provide women with opportunities." In this imagining, the Middle East, the Arab world, and the Muslim world (the three of which he used interchangeably) remain places where misogyny reigns.

Constructions of Taha and Ammash thus put a finer point on hegemonic Western narratives that presume that Arab and Muslim cultures—even at their best—are incapable of fully modernizing. Such a narrative can only lead to a conclusion that fits perfectly with the aims of U.S. empire: Arab and Muslim cultures are unfit to develop advanced scientific knowledge and technology, unless under the supervision and discretion of U.S./Western/Global North powers. Taken to the extreme, this narrative implies that Arabs and Muslims should be prevented from cultivating science and technology altogether.

Conclusion

The war on terror bore at least three Orientalist figurings of the bioterrorist: from the classic, rather caricatured figure of the violent primitive, to the analogized germ, and finally to the semi-modern female scientist-terrorist who is technologically, if not morally, equipped. A presumed cultural pathology of violence threads through all three characterizations, as does the notion that Arabs and Muslims lack the key dimension of civilization—Western morality. Collectively, the three figures illustrate that the "clash of civilizations" worldview endures, painting Arabs and Muslims as infected by a feverish will to spread destruction and as unable to culturally progress. In detailing U.S. constructions of the bioterrorist and their appearance in a range of U.S. sources from the popular press to less accessible realms of national security and science, I have aimed to lay bare their connections to anti-Arab, anti-Muslim, sexist, and colonial Western discourses. Discourse on bioterrorism constitutes yet another site where critical scholars and activists can interrupt the ideologies of the war on terror and their interminably harmful effects.

The malleability of the bioterrorist construction, particularly between the two seemingly disparate figures—the Arab/Muslim primitive and the highly educated Arab/Muslim scientist—indicates the construct's symbolic utility in capturing a range of targets: it ensnares incontrovertibly technologically advanced states like Iraq, but also less equipped, diffuse networks like Al Qaeda, while simultaneously producing a common narrative of both as violent, death-loving, and immoral. The Arab/Muslim scientist-terrorist figure is, I believe, particularly powerful in serving U.S. bio-imperialism due to its subtle reinforcement of the clash of civilizations worldview: the figuring suggests that even the best of Arab and Muslim populations—the potentially feminist, potentially civilizable elite, Western-educated woman—will fail to use the tools of (Western) civilization for anything but terror. The moral of this story is that Arab and Muslim cultures are irreparably backward and evil—a conclusion that then bolsters the foundational rationale for not only continued U.S. (imperial) control over Arab and Muslim cultural and technological development, but, as we shall see in the chapters to come, a wide range of preemptive measures by the United States in the name of "biodefense."

2

From Practicing Safe Science to Keeping Science out of "Dangerous Hands"

The Resurgence of U.S. "Biodefense"

Sadly, one of the insanities of the chase after military security is a world-wide competition in research and development on biological warfare.... These activities are aimed at practicing the large-scale deployment of the most contagious enemies of man that he can discover or invent. Our personal security must then depend on the depth of the technical competence of the men responsible for the research.... We have to be unforgivingly harsh in our judgment of mistakes and leaks.

—U.S. molecular biologist and leading government science adviser Joshua Lederberg, "Congress Should Examine Biological Warfare Tests," March 30, 1968

Our nations' best scientists must support policy-makers in their efforts to make progress toward measures that will counteract the threat from advances in weapons technology that could be misused by governments or as terrorist threat agents.

—Joint statement by the presidents of the U.S. National Academy of Sciences and the Royal Society, the UK national academy of science, "Scientist Support for Biological Weapons Controls," November 8, 2002

The two quotes above—by scientists in different periods—represent the shift in scientists' attitudes toward biology and warfare. In the earlier era, scientists questioned their role in arms buildup and acknowledged their duty to mitigate the hazards of their research. In the post-9/11 era, scientists focused on counteracting terrorist acquisition of weapons technology and not the dangers of the weapons technology itself. This change in focus coincided with significant government investment in the field of biological weapons research during the war on terror—in the name of "biodefense." The latter spawned research into highly pathogenic "biological agents" (any infectious substances such as bacteria, viruses, and fungi, and their associated toxins that can be used as weapons).[1] It also spurred the development of a range of "countermeasures"— prophylactics and treatments such as vaccines and antivirals, as well as surveillance technology. This work took place in government defense labs, but also in university and commercial bioscientific and biomedical facilities.

The post-9/11 revamping of biodefense was a significant departure from the low-level government investment of preceding decades, as the industry had contracted significantly since President Nixon banned the offensive biological weapons program in 1969 (in part due to pressure from scientists like Lederberg, the author of the first quote). New security measures, moreover, accompanied this post-9/11 expansion. Branded "biosecurity," this included legislation and security measures limiting access to high-containment, dangerous research from "misuse": for example, higher security clearances to research on biological agents and censorship of scientific publications so as to avoid revealing information on biological agents to the public.

Like other scientific fields tied to the military, the biodefense industry puts into sharp relief any presumption of science as a realm of abstraction. Carol Cohn in her influential piece "Sex and Death in the Rational World of Defense Intellectuals" (1987) forged a feminist critique of military science, demonstrating how nuclear scientists framed their work in gendered terms—for example, the phallic imagery of the missile, which served to sanitize the threat inherent in their work and distance them from its brutal implications. Science and technology studies scholar Joseph Masco (1999) has exposed the troubling racial consequences of militarized science, for instance the nuclear research industry's exploitation of Latino labor and poisoning of Native American lands. In this chapter I explore how U.S. biodefense reproduces inequities that are national, racial, and gendered—specifically I unpack the multiple discursive threads biodefense rests upon: the Arab/Muslim bioterrorist figure, the "innocent" U.S. scientist, and faith in technoscientific progress.

Policing Others: The United States as Global Watchdog

The United States has dominated international arms control since the end of the Cold War. International regulatory mechanisms are designed to favor global powers via their uneven application as well as in the structure of the laws themselves. As gender and global politics scholar Liz Philipose (2008) has noted, international law has "developed in tandem with imperial needs to justify colonization, slavery, occupation and decision-making authority over the lands and peoples of non-European extraction" and, further, reflects a civilizing project aiming to "sustain the ongoing dichotomy between the civilized and the backward" (104–105). The 1972 international ban on biological weapons, the Biological Weapons Convention (BWC), arose in the context of global powers (notably the United States, the United Kingdom, and France) abandoning biological weapons in favor of nuclear ones, which they both possessed and believed much more effective (Guillemin 2005a). Thus, the protocol banned a weapon global powers presumed they no longer needed, while arming them with a regulatory mechanism to police the biological weapons of other nations. The BWC, coupled with subsequent verification processes authorized by the UN Security Council in the early 1990s, provided the basis for the United States and its allies to control targets like Iraq.

U.S. (and European) control over Iraq had begun in the 1980s when they supplied Iraq with biological and chemical weapons as a strategic move against Iran during the Iran-Iraq War (Barnaby 2000; Central Intelligence Agency 2007; L. Cole 1997). The United States had overlooked Iraq's use of chemical weapons against Iran, even though it violated the Geneva Protocol's ban on the first use of chemical and bacteriological weapons in war (Rapoport 1999).[2] But once Iraq fell out of favor with the United States, leading to the First Gulf War in 1991, the United States and other powerful nations targeted Iraq's weapons programs with the verification protocols of the BWC. Iraq was subject to UN weapons inspections following the war, and again in 2002. These inspections set the stage for further U.S. domination of Iraq: Iraqi biological weapons possession served as a primary rationale for a second invasion of the country in 2003.

The pro-war propaganda leading up to the invasion of Iraq in March 2003 maligned Iraq as flouting international law—the United States accused Iraq of possessing biological weapons (and other WMDs). President Bush, members of the Defense Department, and high-level weapons experts, as well as mass media and other pundits, vigorously painted the specter of Iraq as embodying the threat of bioterrorism. Iraq and its leader, Saddam Hussein, had been demonized as an Arab/Muslim Other since the First Gulf War, and in the post-9/11 context U.S. pundits easily floated connections between Iraq and Al Qaeda, WMDs, terrorism, and the anthrax attacks.[3] While most of these claims proved baseless, and were later revealed to be either exaggerated or altogether false (Center for Public Integrity 2008; Kimball 2003; Pitt 2002), in the lead-up to the invasion their symbolic utility proved undeniable.

On February 5, 2003, Secretary of State Colin Powell gave a speech to the UN Security Council about Iraqi biological weapons. He showed pictures of mobile research laboratories and held up a small container of white talcum powder (figure 2) meant to simulate dry anthrax (Vogel 2012), stating, "Less than a teaspoon full of dry anthrax in an envelope shut down the United States Senate in the fall of 2001," followed by "Saddam Hussein could have produced 25,000 liters. If concentrated into this dry form, this

FIG. 2. U.S. Secretary of State Colin Powell speaking to the United Nations, February 5, 2003. (Reuters/Ray Stubblebine.)

amount would be enough to fill tens upon tens upon tens of thousands of teaspoons" (Powell 2003). By visually and rhetorically reducing Iraqi bioweapons capability to anthrax powder, Powell drew on the gripping image of the Arab/Muslim Other as menacing germ. Like the figure of the suicide infector I described in the previous chapter, this germ figure further stoked the construction of Iraq as a potent biothreat, and state officials used it to rally action.

This threat imaginary about Iraq bolstered U.S. rationale for invasion, reinforcing the United States' ostensible role as global police of the biological weapons capacity of nations around the world. The image of Colin Powell also represented U.S. exceptionalist narratives during the period: pictorial and rhetorical invocations of high-ranking people of color in the Bush administration, particularly Secretary of State Powell and his successor, Secretary of State Condoleezza Rice, served in the political theater of the war on terror to present the United States—by virtue of its multiculturalism—as democratic and fair, a land of equality and justice.[4] Like discursive gestures to morality, civilizational advancement, and feminism described in the previous chapter, these depictions rendered U.S. actions domestically and abroad as serving these lofty, benevolent ideals.

The horrific effects of U.S. empire cannot be overstated. Consider the catalogue of topics about the U.S. invasion of Iraq in the anthology *Feminism and War: Confronting U.S. Imperialism* (Riley, Mohanty, and Pratt 2008): record numbers of civilian casualties; the body count in general; racialized, sexualized torture; enlistment of U.S. women into the dangerous culture of gendered violence in the U.S. military; an increase in domestic and sexual violence against women and children worldwide; not to mention the decimation of the economies and infrastructure of targeted countries and the cuts to U.S. social services and the effects on marginalized populations within the United States. In this context, U.S. ability to determine the course of biological warfare globally constitutes cause for great concern—it is one more node in a constellation of U.S. power augmented during the war on terror.

Building Up Domestic Capacity: The "Defense" Moniker

While the U.S. national security state has aggressively suppressed the capacity of other nations, it has shielded its own programs from scrutiny. This was pointedly illustrated in February 2003 when a Canadian-based coalition attempted to apply the BWC to the United States. This delegation—comprising parliamentarians, scientists, academics, and religious and union leaders from Canada, Europe, and the United States—showed up to inspect a U.S. Pentagon facility suspected of developing and stockpiling chemical and biological weapons. They were unceremoniously turned away.[5] The U.S. media barely covered the incident.[6] One journalist labeled it a "spoof" (E. Rosenberg 2003).

In contrast to the considerable public outcry against U.S. engagement in Iraq and international weapons policing, there has been far less criticism of the United States' own weapons arsenal. This discussion has mainly taken place among specialists—scientists and bioweapons watchdog groups—and a handful of journalists. Indeed, the U.S. national security state has such a tight hold on the arms control arena that it can exert near-unilateral authority globally, while maintaining secrecy around its own programs.

The dismantling of the U.S. biological program under Nixon had never been complete. Several military and CIA-run labs retained stocks of lethal bioweapons, including anthrax, despite Nixon's order that they be destroyed (Falk 1990; Guillemin 2005a). In addition to outright retention of prohibited materials, the United States maintained its offensive capacity through the research it conducted on the effects of biological agents and their delivery mechanisms. Moreover, the U.S. practice of open-air testing of biological weapons in remote locations to assess their effects on animals and plants continued at least until 1973, along with testing of simulants to garner information about the potential geographic spread of an attack (Barnaby 2000; L. Cole 2016).

Former national security adviser Henry Kissinger outlined the U.S. approach toward defensive research in the wake of Nixon's ban, stating in National Security Decision Memorandum 35 that "research and development for defensive purposes does . . . not preclude research into those offensive aspects of bacteriological/biological agents necessary to determine what defensive measures are required." Kissinger's statement indicated the United States' lack of commitment to truly ending its offensive capabilities, but also signaled the inherently blurry line between offensive and defensive programs. The production of vaccines, antivirals, and other treatments against infections caused by germ weapons—a mainstay of defensive programs—hinges on the possession and production of biological agents. Many vaccines, for instance, comprised inactivated forms of the pathogens themselves, such that acquiring and growing these pathogens—and in sufficient amounts—become necessary to produce enough vaccine for the target population.[7]

The codification of the BWC protocol three years later, in 1972, offered no clarity on how signatory states should navigate the line between offensive and defensive programs. In fact, the BWC implied that states could produce or develop biological agents if justifiable for "prophylactic, protective or other peaceful purposes."[8] Given the ambiguity between offensive and defensive research, Susan Wright and Stuart Ketcham (1990) have argued that key metrics of a biological program's offensive capacity are the size and scope of its support systems, namely, infrastructure for production and delivery of biological agents as well as communication and command systems.[9] Under Nixon, then, the United States did not entirely give up its ability to wage biowarfare, but did drastically reduce its arsenal. President Carter further diminished the program.

With the election of Ronald Reagan to the presidency, U.S. interests and war strategy shifted. Amid concerns about potential Soviet capabilities, Reagan renewed attention to biological weapons, as well as chemical and nuclear ones. The period from 1981 to 1984 experienced an upsurge under a Department of Defense (DOD) program known as the Biological Defense Research Program. The program pushed further into offensive territory, expanding the number of biological agents under research and producing putative Soviet bioweapons for the alleged purpose of testing them against U.S. protective systems (Piller and Yamamoto 1990). In fact, the DOD publicly revealed that it was developing "deterrents," that is, offensive capabilities meant to deter enemies from first use (Huxsoll, Parrott, and Patrick 1989; Quigley 1992; S. Wright 1989).

The end of the Cold War ushered in U.S. status as sole superpower and a brief respite from the country's biological arms buildup. However, in the late 1990s biowarfare specialists persuaded President Clinton that, despite the dissolution of the Soviet Union, the Soviet biological program remained a threat. Earlier in the decade Russian defectors had alleged that the Soviet Union engaged in offensive bioweapons research—and on a massive scale—although Russian leaders never definitively admitted to this.[10] Many U.S. officials seemed to believe the allegations and expressed concern that the components of the former Soviet program had made their way to other nations— raising the specter of offensive bioweapons capabilities outside of U.S. control.

The U.S. response was to revisit Cold War escalation strategy and ramp up its own biological program. In 1998, Clinton inaugurated a new biodefense effort, a "civilian biodefense" program (the Bioterrorism Preparedness and Response Program) run by the CDC and comprising nonmilitary and nongovernmental domains, namely, crisis management and public health domains.[11] Military sectors also began research efforts—believed to have been largely kept secret from the Clinton administration—that tread the defensive-offensive line. In the Clear Vision Project, the CIA contracted with Battelle Memorial Institute to reverse engineer Soviet biological bombs (made to disseminate biological agents) and in doing so, according to a number of legal scholars, violated the BWC's ban on offensive biological weapons research (Miller, Broad, and Engelberg 2001; Tucker 2004). Defenders of the research argued that it provided important information for threat characterization and to test countermeasures.

Project Bacchus, run by the DOD, entailed the building of a mock biowarfare factory to produce simulants of anthrax for testing biological weapons production scenarios. Like Clear Vision's reproduction of Soviet biological weapons, the building of germ dissemination capabilities provoked questions about whether it could be considered solely defensive (Miller, Engelberg, and Broad 2001). What is indisputable is that U.S. engagement with such murky defensive-offensive research, as well as its expansion of biodefense into civilian sectors, boosted U.S. strategic advantage in bioweapons.

The post-9/11 augmentation of U.S. geopolitical power further expanded biological warfare research and development, and the United States continued to produce dangerous pathogens in the hopes of advancing knowledge and to produce effective countermeasures. Of the dramatically increased biodefense

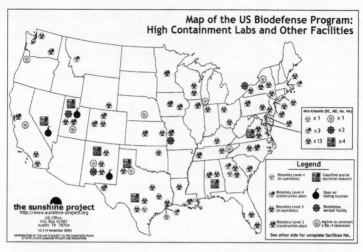

FIG. 3. Map of existing and proposed high-containment U.S. biodefense labs (biosafety levels 3 and 4). November 4, 2004. (Reprinted with permission from the Sunshine Project.)

budget (from $1.4 billion in fiscal year 2001 to about $5 billion in 2002, and over $7 billion in 2004), funding was primarily earmarked for research, development, and acquisition of medical countermeasures such as vaccines and antivirals.[12] This was accompanied by a mushrooming of new high-containment laboratories across the nation that could conduct research involving dangerous biological agents like anthrax and smallpox.

Laboratories that are designed to enclose infectious agents are designated with Biological Safety Levels (BSL) set from 1 through 4 (with 4 pertaining to the most dangerous pathogens, such as smallpox or Ebola); they house specialized equipment and require scientists to follow specific safety protocols and best practices. In 2002 the Department of Homeland Security spent $70 million to open more of these labs (Guillemin 2005a). The Sunshine Project, an international bioweapons watchdog NGO, took a snapshot in November 2004 of the national array of high-containment labs—specifically BSL-3 and BSL-4 labs for containing lethal pathogens. The map (figure 3) showed existing as well as planned labs: the former designated with a biohazard sign, the latter designated with a biohazard sign plus a dot in the sign's upper-right corner (Sunshine Project 2004).

This proliferation of biodefense continued to be outsourced to the civilian sector (mainly the Department of Health and Human Services [HHS] agencies) as well as private industry. The National Institutes of Health (NIH)—the research agency of HHS—was funded to conduct research and development for new biomedical countermeasures at over $1.5 billion per year in fiscal year 2003—thirty times the investment in fiscal year 2001 (Bush 2004a).[13] The pharmaceuticals industry—the face of neoliberal capitalism in the biomedical sphere—played an increasingly large role. The government enlisted this large and powerful congressional lobby[14] to produce vaccines and other drugs against biological agents by offering incentives such as tax breaks and indemnities (since vaccines are not the most lucrative of pharmaceutical products).[15]

This reliance of government on private industry, begun under the Clinton administration, has reduced compliance monitoring. Since the late 1990s, the U.S. state has sought to decrease U.S. transparency in international weapons compliance. In particular, the U.S. Congress and the pharmaceutical industry, arguing that transparency through onsite inspections of military and pharmaceutical facilities would endanger commercial proprietary and national security interests, pressured Clinton into weakening the treaty's inspections regime at the fourth review of the BWC in 1997[16] (Guillemin 2005b; S. Wright 2002). The Bush administration unabashedly maintained this corporate biodefense policy, rejecting the ratification of a strengthened international bioweapons treaty during its quintennial review in 2001, and thereby halting the verification protocol. Once again, the United States cited national security and commercial interests (Dando 2006; Findlay 2006). In this way, the profit imperative of the private sector worked hand in hand with U.S. imperialism, further enabling what some have termed the "biodefense industrial complex" (Fidler and Gostin 2007, 148). Empire and capitalism easily superseded international disarmament aims, enabling the U.S. state to shrug off substantive limits to its industry's activities.

The U.S. push for international biological disarmament in the late 1960s had been the product of both concern about the morality of biological weapons and strategic disinterest given the rise of nuclear weapons (the latter's proven power versus the former's questionable efficacy). Both considerations hinged on the notion that some nations and peoples are unfit to possess weapons. In February 1967, leading scientists penned a letter to the White House opposing the use of biological weapons by invoking the notion of good/bad national governments: "CB [Chemical[17] and biological] weapons have the potential of inflicting, especially on civilians, enormous devastation and death which may be unpredictable in scope and intensity; they could become far cheaper and easier to produce than nuclear weapons, thereby placing great mass destructive power within reach of nations not now possessing it; they could lend themselves to use by *leadership that may be desperate, irresponsible or unscrupulous*" (quoted in Robin Clarke's 1968 *The Silent Weapons*; emphasis added).

The notion that some national governments were unfit to responsibly manage dangerous weapons had been, since the post–World War II period of rising U.S. power, a continuous thread in narratives of U.S. exceptionalism—namely, of the United States as global steward acting with just, altruistic motives. The trope of the ill-intentioned, unfit Other was its counterpoint—a trope the United States has used to demonize entities that threaten U.S. power: during the Cold War, it was applied to the Soviet Union and communism, shifting at the end of the Cold War to "terrorists" and "rogue states" (a history I described in the introduction). In the context of biological weapons, the trope of the unfit Other proved useful not only in the U.S. push for the ban on offensive weapons in the late 1960s, but also in subsequent periods of renewed U.S. interest in biological arms as "defense."

In this vein, the post-9/11 national security state mobilized the era's most well-worn trope of the Other—the Arab/Muslim terrorist—to enhance U.S. biodefense. Section 1013 of the USA PATRIOT Act, which focused specifically on biological warfare, outlined an expanded role for public health (the subject of the next chapter) and the bioscientific research enterprise in bioterrorism preparedness and response, authorizing funding allocations accordingly.[18] The act cited as rationale Osama bin Laden's statements of his intent to acquire WMDs: "public pronouncements by Osama bin Laden that it is his religious duty to acquire weapons of mass destruction, including chemical and biological weapons." The section went on to describe a "callous disregard for innocent human life as demonstrated by the terrorists' attacks of September 11, 2001," and "the resources and motivation of known terrorists and their sponsors and supporters to use biological warfare." The characterization of bin Laden and the September 11 attackers invoked the figure of Arab/Muslim Other I described in the previous chapter—defined by cultural and moral backwardness, and thus uniquely motivated to harm innocents. The law's invocation of this pervasive characterization thus helped concretize a bioattack as imminent and made it a basis for U.S. state action (i.e., to build up its biodefense industry).

On April 28, 2004, President George Bush issued Homeland Security Presidential Directive (HSPD) 10, "Biodefense for the 21st Century," calling for renewed attention to biodefense. The nine-page document began by quoting Bush's earlier (February 11) remarks on the importance of countering international weapons proliferation: "Armed with a single vial of a biological agent, small groups of fanatics, or failing states, could gain the power to threaten *great nations*, threaten the world peace. America, and the entire civilized world, will face this threat for decades to come. We must confront the danger with open eyes, and unbending purpose" (Bush 2004b; emphasis added). The passage constructed a divide: malevolent groups in possession of biological agents and ready to wield them, against "great nations" such as the United States who are "civilized" and peaceful. Not only does the latter group's possession of biological agents go unmentioned, but by implication, their possession is presumed to be for defensive purposes only.

The U.S. national security apparatus further couched its bio-defense industry in exceptionalist terms by recuperating the notion of "dual use." Dual use originated as a Cold War term pertaining to a nuclear-focused quandary: that scientific research could be applied toward either civilian or military purposes (Atlas and Dando 2006; Chyba 2002). In post-9/11 rhetoric, however, the term's usage shifted from a distinction of domains to that of intent: "dual use" came to mark certain research materials—those involving biological agents, for instance—as having the potential for both good and evil purposes. In 2004, HHS set up the National Science Advisory Board for Biosecurity (NSABB) to create rules for "dual use" research; the official press release announcing the formation of the board stated: "HHS Secretary Tommy G. Thompson today announced that HHS will lead a government-wide effort to put in place improved biosecurity measures for classes of *legitimate biological research that could be misused* to threaten public health or national security—so-called 'dual use' research" ("HHS Will Lead" 2004; emphasis added). The distinction being made here was between "legitimate research" (for either health or military/security purposes) by scientists in the United States, and its misuse—by implication the bioterrorist Other. This distinction thus legitimated research according to *who* was conducting it, the assumption being that U.S. science fields were motivated by good intent.

In another section of the Biosecurity Board's press release, a quote from Secretary Thompson elaborated his belief in U.S. exceptionalism: "*Our nation* has been a world leader in life sciences research because of our emphasis on the importance of the free flow of scientific inquiry. Yet, sadly, *the very same tools developed to better the health and condition of humankind can also be used for its destruction*" ("HHS Will Lead" 2004; emphasis added). Here, U.S. research is further marked as globally well intentioned—it seeks to better "humankind," further legitimizing U.S. research as necessary to protect the world against the inherently suspect actions of those deemed Other.

The Nuclear Threat Initiative, an NGO that works closely with the government on WMD threats, published an article on international arms control that explicitly highlighted this turn to marking good/bad intent: "Since the BWC prohibits the possession of biological agents for offensive military ends while permitting their use for peaceful scientific, therapeutic, or defensive purposes, judgments of treaty compliance may hinge on an *assessment of intent*" (Nonproliferation Studies 2004; emphasis added). The focus on intent completely bypassed an assessment of whether the research itself qualified as offensive or defensive—an important, if admittedly blurry, line. This too effectively shifted the focus from considering the research itself to *who* undertakes it. In the era of U.S. superpower status, government officials easily enacted these frames to bolster de facto U.S. escalation of research activities, coding them as not only defensive but good for the world.

As the racialized bioterrorist Other came to be the essential building block in the logic of U.S. domestic buildup, it had increasingly dangerous implications for those occupying the status of Other, especially Arabs and Muslims working in U.S. defense fields. The plight of Arabs and Muslims in U.S. defense fields mirrored the escalation of institutional and societal discrimination during the war on terror.[19] Ayaad Assaad, a physiologist and veterinarian and an Egyptian American, worked at Fort Detrick for nearly a decade before being let go in 1997 during a round of staff cuts. In 2001, an anonymous letter implicated him in the anthrax mailings—and this was only the latest antagonism he faced. Assaad had battled for years against the hostile environment for Arab Americans in defense science, finally filing a discrimination lawsuit; the investigation that ensued would reveal internal Army documents detailing rampant racial and racialized sexual harassment (Tuohy and Dolan 2001; Warrick 2002; Weiss and Warrick 2002).

Assaad's treatment was in marked contrast with that of white male scientist Bruce Ivins, whom the FBI identified as the perpetrator of the 2001 anthrax mailings. The U.S. government and corporate media responded to the identification of Ivins with attempts to establish his good intent. He was depicted as a well-intentioned mad scientist rather than a violent white male terrorist (D'Arcangelis 2015). This repositioning of Ivins as a scientist rather than a bioterrorist reestablished the stark boundary between the two identities and reflected its fundamentally racial contours. Accordingly, Assaad's status as scientist—mediated by security discourses that marked him as threatening Other—was always treated as suspect.

The Mask of Biosecurity: Externalizing Danger

U.S. intensification of its bio-imperial capacity harms not only its intended targets. All bioscientific research entails risks such as laboratory accidents, unintentional leakages, and transport mistakes. Research on dangerous pathogens, also subject to these mishaps, thus yields significant biohazards to laboratory workers and those in the geographic vicinity of the labs.

In 2004, three laboratory workers in a Boston University biosafety level 2 lab, working with what they mistakenly thought was a benign strain of tularemia, became ill with a virulent and potentially fatal strain. That same year, seven researchers at a biosafety level 2 lab in Oakland Children's Hospital were exposed to live—instead of dead—anthrax, sent by mistake by a vendor in Fort Detrick, Maryland (Hecht and MacKenzie 2005). Other notable accidental releases and exposures that took place in the early twenty-first century have been catalogued by the Sunshine Project. In 2006, the organization updated the map (figure 3) to not only include the new labs built in the meantime, but also show the releases that had occurred from 2001 to 2005.[20]

These lab accidents and leakages are not historically anomalous. During the height of the active U.S. program in the 1950s and 1960s, at Camp Detrick, the former hub of military research on biological weapons, there were 3,330 accidents between 1954 and 1962; half involved lab personnel, of whom 77 percent were infected (Hersh 1968).[21] The dire consequences of these accidents have best been demonstrated by the Sverdlosk disaster in 1979, when weaponized anthrax leaked from the Soviet bioweapons laboratory, killing an estimated one hundred people (Guillemin 1999; Inglesby et al. 2002).

The revamped post-9/11 biodefense industry continued this trend. The CDC has recorded about twenty accident reports of infectious pathogens a year since 2004—a number researchers think is probably underreported, and which increased to thirty-two in 2007 (Kaiser 2007). The growing research on putative biological agents likely increased the potential for these mishaps—the number of people gaining clearance to work reached about 20,000 at 400 sites around the United States by 2007, ten times more than before 9/11 (MacKenzie 2007).

I have discussed how the specter of the mal-intentioned Other has enabled the U.S. state to justify domestic buildup. Here I also argue that it helped the state to mask the risky research and laboratory hazards the biodefense industry generates. Focus on the bioterrorist figure pushes aside the ongoing need to assess the prudence of conducting such research. This discursive move has been well described in other arenas of U.S. national security, suggesting a broader pattern. Anthropologist Joseph Masco (1999) has highlighted how, in U.S. nuclear research culture, the post–World War II national security state turned its focus to "national security" threats in ways that both produced and concealed "national sacrifices" (204). Masco describes the building of Los Alamos National Laboratory (a U.S. nuclear weapons complex in Los Alamos, New Mexico) as exploiting Native American lands, producing nuclear waste and environmental contamination, and exacerbating the exploitation of Nuevomexicano labor. The U.S. national security apparatus frames the harms it produces as the justifiable cost of achieving national defense.

Post-9/11 institutionalization of "biosecurity" further enshrouded the "sacrifices" of biodefense research. The biosecurity regime focused on guarding and restricting access to research on dangerous pathogens.[22] This meant tightened security at facilities, increased screening of lab workers, and restrictions on publication of research findings and collaboration. Operationalizing biosecurity focused attention on outsiders obtaining the dangerous research materials, rather than on concerns about internally generated lab hazards.

The new access restrictions, moreover, aimed at specific categories of people. The USA PATRIOT Act had extended post–Cold War era discourse on terrorism to mark both communist countries and countries in the Middle East—whether individuals, groups, or nation-states—as severe threats to U.S. national security. The act also contained a section regulating bioscience, aiming to more tightly restrict access to and transfer of biological materials deemed hazardous (so-called select agents) from categories of "restricted persons."[23] To section 817, "Expansion of the Biological Weapons Statute," it added subsection 175b, "Possession by Restricted Persons." This subsection restricted access to select agents by anyone who was an "alien illegally or unlawfully in the United States" or "an alien (other than an alien lawfully admitted for permanent residence) who is a national of a country as to which the Secretary of State . . . has made a determination (that remains in effect) that such country has repeatedly provided support for acts of international terrorism." The latter category denoted foreign nationals from countries designated by the State Department as "state sponsors of terrorism," which at the time included Cuba, Iran, Iraq, Libya, North Korea, Sudan, and Syria.[24] The restriction's basis in national origin—specifically, to countries identified as hostile to the United States—racialized them via the discourse of terrorism.[25]

In addition to these racialized restrictions based on citizenship status and national origin, the act marked other categories of persons as national security threats: anyone who has been convicted of a crime for a term exceeding one year, fugitives from justice, users of controlled substances, anyone who has been discharged dishonorably from the U.S. Armed Services, and anyone deemed a "mental defective." The delineation of this wider array of Others deemed unfit to access biological agents demonstrates the logics of state targeting. The state marginalizes various abject groups, at different times and places, and to different degrees depending on the dictates of biopower.[26] The discursive gesturing to those the state abandons—the Arab/Muslim Other, the convict, the drug user, the defector, and the disabled—in the name of protecting "life" served to bolster the logics of the biosecurity regime and its aim to keep pathogen research only in the hands of the state and its proxies—in this case high-level biodefense insiders.

The U.S. state did not limit itself to these categories. Steve Kurtz was not part of any of the listed exclusions—he was a white male and had no connections to incarceration, drug use, wartime defection, or disability. He was, however, a lab outsider—a bio-art activist doing critical work about the politics of science[27]—who got caught up in the post-9/11 counterterror regime through an unfortunate series of events surrounding a heartbreaking personal tragedy.

Hope Kurtz had fallen ill and died suddenly one night. Her husband Steve called the police, who then called in the FBI after seeing that his home contained a biology setup and a printed invitation that contained some Arabic writing. While these were all preparatory materials for an upcoming bio-art installation involving high school grade bacteria, the authorities viewed Kurtz as an object of suspicion. What ensued was a farce of scrutiny and punishment against Kurtz and other bio-artists affiliated with him.[28] Kurtz was charged under the PATRIOT Act for possessing a biological agent for a purpose other than "prophylactic, protective bona fide research toward educational or other peaceful purposes."[29] Even after the bacteria was determined harmless, he was indicted for mail and wire fraud for obtaining the samples. Many speculate that his targeting was to establish a test case for the newly expanded legal provisions. It took four years, during which time the FBI investigation pulled in many of his art associates, for his case to finally be dismissed (da Costa 2010; Hirsch 2005; Kane and Greenhill 2007). It was, without question, an extreme response that caused a great deal of harm. It also signified the lengths to which the U.S. state would go to implement its biosecurity regime.

It was not only the much-maligned PATRIOT Act that invoked familiar conceits of Othering to enact biosecurity. Alongside the codification of the ramp-up of research, the Public Health Security and Bioterrorism Preparedness and Response Act of 2002 devoted vast resources to producing biomedical countermeasures, disease monitoring, and surveillance infrastructure, as well as enhancing control over biological materials. It outlined new security measures to "prevent access to such agents and toxins for use in domestic or international terrorism or for any other criminal purpose."[30] These included registration requirements for individuals to possess, use, or transfer biological agents, and tracking the relationship between registered persons and their collaborators, and more severe penalties for registration violations.

The Bioterrorism Preparedness Act signaled the institutionalization of biosecurity in another way—by eclipsing existing discourses of biosafety. "Biosafety" is the term for laboratory guidelines aimed at minimizing the release of and exposure to dangerous research materials—that is, the spills, leakages, and other accidents outlined above. Biosafety guidelines outline proper laboratory containment and handling of biotechnology hazards related to pathogens and two other categories of hazardous lab material—genetically modified organisms (GMOs) and GMO pathogens (Sunshine Project 2003; WHO 2005b).

Biosafety ended up being only a brief part of the Bioterrorism Preparedness Act. A few short lines in section 351A ("Enhanced Control of Dangerous Biological Agents and Toxins") describe measures to ensure "proper training and appropriate skills to handle such agents and toxins; and proper laboratory facilities to contain and dispose of such agents and toxins," and later "procedures to protect the public safety in the event of a transfer or potential transfer of such an agent or toxin in violation of the safety procedures." Yet, section after section is devoted to biosecurity—how to prevent access to terrorists, registration of individuals working with dangerous agents and toxins, and procedures for the attorney general to vet these names. With only minimal mention of biosafety in the act, the biosecurity regime largely displaced attention to protections against the hazards of biological research.

These legal measures forced change upon the institutional cultures of science, which were encouraged to safeguard their research from the bioterrorist Other. The National Research Council of the National Academy of Sciences, which works closely with the federal government, reflected this shift in its 147-page report entitled *Biotechnology Research in an Age of Terrorism*. The committee's stated goal was to "consider ways to minimize threats from biological warfare and bioterrorism without hindering the progress of biotechnology, which is essential for the health of the nation . . . because almost all biotechnology can be subverted for *misuse by hostile individuals or nations*" (National Research Council 2004; emphasis added).

As national scientific organizations took up this mantle of biosecurity, they implemented guidelines that balanced attention to biosecurity with the dominant interests and practices of the sciences—namely, the transparency and collaboration essential to the scientific research process and information dissemination (Bhattacharya 2003; Gaudioso and Salerno 2004; Kwik et al. 2003).[31] In January 2003 a group of science journal editors published a statement in *Nature Medicine* that advocated vetting papers for content that might be used for bioterrorism and modifying them accordingly ("Editors' Statement" 2003). Many scientists responded with concern about the possible impacts of the new security guidelines on their scientific productivity and the costs of doing research (Dias et al. 2010; Kaiser 2005). They were not, then, overly critical of the aims of the regime or the racialized discourse that was its lynchpin, and many scientists shifted their research to fit the biodefense agenda so that they could obtain research funding (Gellene 2003). Scientists who undertook this shift may not have actively supported the biodefense agenda, but their actions helped enact it.

The adoption of the regime of biosecurity was a significant departure from older cultures of scientist accountability. As the epigraphs at the beginning of the chapter suggest, scientists of the 1960s were duly concerned with the biohazards of research on pathogens as well as with scientist participation in the biowarfare industry. In the post-9/11 era, no longer were scientists arguing for U.S. disarmament, but instead for enhanced security measures to safeguard their research. In largely accepting biosecurity discourse, the bioscientific community enabled a perspectival change—to viewing danger as primarily emanating from outside the laboratories, rather than primarily internally generated. From this standpoint, U.S. scientists are innocent, unwitting victims of terrorism, which serves to further divert scrutiny from the role bioscientific research and bioscientists play in biowarfare.

Safeguarding Technoscience

The U.S. biodefense scientist is the figure upon which falls not only the presumption of peaceful, good intent, but also technoscientific authority. Unfettered scientific inquiry and technological advancement, it is assumed, can and will lead to the betterment of humankind—the only caveat being that it must be kept out of the hands of the unfit, ill-intentioned Other. The logic of biosecurity, then, is to safeguard the ability of the scientist to press forward in this onward march of research.

In 2001, Australian scientists reported that they had inadvertently created a new, highly virulent mousepox strain (a pox virus that is a relative of smallpox, but whose primary host is mice rather than humans). They created this strain while conducting research in 1998–1999 to engineer a mousepox virus that would act as a contraceptive to overpopulating mice. The strain was so deadly that it killed animals that had been vaccinated. As news of the experiment circulated to the U.S. biodefense community, many raised the "dual use" specter—knowledge of how to engineer lethal vaccine-resistant strains was now publicly available for "misuse." Amid the ensuing discussion about whether scientific journals should censor what they publish (the previously mentioned statement of science journal editors was one result of the discussion), many pushed for more research of this kind—to create and study other similarly lethal pox viruses. The risk, as already mentioned, lies in the possibility of these new, highly lethal life forms leaking out and causing—to a much greater degree than non-enhanced pathogens—illness in nearby locales, or worse, starting an epidemic. Yet, proponents of the research argued that it would help the United States prepare for, and even pre-empt, the knowledge that a putative bioterrorist would garner if they were to achieve re-creation of pox viruses (Aldhous and Reilly 2006).[32] Such an approach reflected conviction in the necessity of scientific advancement as well as faith in biosecurity—that further experimentation with the virulent strain would yield benefit for U.S. biodefense and could be contained within the industry's control.

More than half of the 2004 budget (nearly $3 billion) the Bush administration allocated the National Institute of Allergies and Infectious Diseases to produce countermeasures went to the development of new countermeasures. This meant that instead of just adding to existing stockpiles of vaccines and antivirals for known pathogens, researchers conducted experiments to develop new treatments for these pathogens—sometimes through genetic engineering of more toxic and drug-resistant strains (Center for Counterproliferation Research 2003).

In 2005, researchers re-created the long-dead 1918 pandemic flu strain (at which point it was labeled by the CDC as a "select agent" due to its lethality). Subsequent debates in the scientific community and the press focused almost entirely on weighing the benefits against the risks in terms of the research's dual use implications, in this case the accessibility to "terrorists" of the sequencing information, if openly published (Aldhous and Reilly 2006; Sharp 2005). Unlike in the 2001 mousepox incident, re-creation of the flu strain was intentional—not accidental. The regime of biosecurity had taken center stage, guaranteeing the headlong rush into perilous biodefense research.

Post-9/11 acceleration of such risky research reveals the logic of state protectionism in U.S. biological warfare research. Like other militarized sciences, the industry embodies a protectionism infused with white masculinity and technoscientific authority.[33] I have in my previous work on U.S. biodefense argued that it is a form of institutionalized "white scientific masculinity": institutions of science and security associated with white males (e.g., defense sciences) get construed as infallible founts of knowledge and expertise.[34] The post-9/11 regime of biosecurity thus garnered the U.S. biodefense industry, a trusted institution of "white scientific masculinity," an even greater ability to conduct risky research without significant scrutiny.

But no case better illustrates U.S. impudence in pursuing new avenues of risky research than that conducted with smallpox. Smallpox is known for its historical devastation—it has an approximately one-third mortality rate and was responsible for almost half a billion deaths worldwide in the twentieth century. It wreaked havoc in Native America centuries ago when European settler-colonists—in an early example of biological warfare—intentionally infected native populations who, without prior exposure, were decimated by the disease (Christopher et al. 1997; Duffy 2002; USAMRIID 2004). In 1979, smallpox became the first disease to be successfully eradicated from the world.[35]

Neither the warfare genealogy of smallpox—as a deplorable tool Europeans had used to commit Native American genocide—nor its status as an internationally eradicated disease and thus a minimal threat—seemed to influence the resurgence of attention the United States gave smallpox at the end of the Cold War. In the context of the breakup of the Soviet Union, U.S. national security circles feared its deployment by so-called terrorist networks and rogue states. The triumph of eradication, moreover, had birthed a new dilemma: whether stocks of variola (the virus that causes smallpox disease) existed outside of World Health Organization (WHO) knowledge, and thus whether it should retain a few guarded stocks for prophylactic, research purposes.[36] The WHO would allow the United States and the former Soviet Union to hold onto stocks of variola—for research purposes—in their high-containment labs (the CDC and the Russian Research Institute for Viral Preparations, respectively).[37] Thus, the public health triumph of smallpox eradication contorted, within the U.S. national security community, into a problem of U.S. vulnerability—a population with little immunity to smallpox and thus highly susceptible to a smallpox bioweapons attack.

The post-9/11 era drastically escalated this post-eradication specter, painting Iraq in particular as a looming threat.[38] This was extraordinarily ironic since the only military deployment of smallpox has been by Europeans against Native Americans (not to mention the false victimhood the United States adopted to justify the 2003 invasion of Iraq). This incongruity notwithstanding, the United States embarked on hazardous research on the variola virus—namely, genetic engineering of novel strains. These are strains to which—by virtue of their novelty—humans would have little or no immunity. As with other instances of pathogen proliferation in the name of biodefense, the strategy was to prepare for and pre-empt others who may be doing the same thing. In 2004, the United States put forward a request to the WHO, in charge of overseeing smallpox studies, for approval to conduct genetic engineering experiments with variola, including creating cross-species hybrids by splicing variola genes into the genomes of other orthopoxviruses; if approved, it would change the 1994 WHO guideline banning such work (WHO 2004).

This request was considered at the World Health Assembly meeting in May 2005. It met with substantial opposition from many countries expressing concern about this escalation of research and potential accidental escapes. Nevertheless, the WHO approved all but the most extreme genetic engineering activities (i.e., the splicing of variola genes into other orthopoxviruses) (Enserink 2005). In 2007, the WHO finally banned all research involving genetic engineering of the variola virus, after pushback by many member countries and NGOs, particularly from the Global South, including the Sunshine Project and the Third World Network, a transnational development and environmental policy NGO based in Malaysia (Third World Network 2007b).

The U.S. push for genetic engineering highlighted once again a faith in technoscience—as long as it remained under the purview of U.S. state biosecurity. The United States, moreover, relaxed its overall restraints on smallpox research in the post-9/11 climate—by continuing to postpone destroying variola stocks. Soon after the WHO agreed to allow the United States and the former USSR to hold onto stocks of the variola virus for research purposes in 1984, it had recommended their destruction once sufficient information had been garnered. In 1990, the WHO assessed that enough research had been conducted (e.g., genome sequencing and clone fragment libraries[39]) and set a date for the destruction of the remaining stocks—1993. Yet, when that time came, the WHO postponed destruction to 1995 due primarily to pressure from the United States and Russia, both of which argued that they still needed the stocks for further research and to develop new countermeasures. Proponents of postponement even cited the achievement of full genome sequencing—intended to obviate the need for preserving viral samples—as presenting a new danger: scientists could use this to re-create the variola virus (Mahy et al. 1993; Hammond 2007). Technoscientific advancement had generated more risk, yet remained the mainstay of U.S. decision making in its pursuit to control variola.

Over the years many nations and independent organizations internationally as well as within the United States have argued for the destruction of these remaining stocks to stem potential proliferation of variola. But in the post-9/11 climate, the U.S. biodefense community has only further entrenched its position to retain the stocks in the name of national security. In 2002, after two subsequent postponements, destruction of the stocks was postponed indefinitely (Third World Network 2007b; WHO 1999).

The biosecurity regime has been successful in shifting attention away from the dangers that the research industry internally generates—in the form of lab accidents, leaks, and other mistakes, as well as mal-intended deployment by industry insiders like Bruce Ivins, the white male biodefense scientist who was the FBI's final suspect in the anthrax mailings of 2001. It is precisely the buildup of domestic U.S. biological warfare capacity that poses a tangible, grave danger, domestically and internationally, to those intentionally targeted by the U.S. national security regime—Arabs, Muslims, and other negatively racialized groups and nations—as well as those affected by the biohazards of the research process itself—lab workers and others in physical proximity of dangerous biological agents. That the buildup continues despite these severe costs attests to the power exerted by U.S. bio-imperialism and the national security state and industry interests that drive it.

Conclusion: The Voice of Scientists

As the foot soldiers of the biosecurity regime, scientists were key to its implementation. Unlike scientist attitude in the pre-1969 ban era (mentioned in previous sections of the chapter), most scientists of the post-9/11 era appeared much less concerned with the dangers of biological research or their role in warfare, and more concerned with protecting research from interference from new government rules. Thomas Butler, an infectious disease medical researcher and a white male, was the first scientist to be tried for biosecurity offenses after losing thirty vials of plague bacteria. He was found guilty in December 2003 of several charges under the Public Health Security and Bioterrorism Preparedness and Response Act of 2002, including shipping plague without government permission. Butler was largely seen in the scientific community as a casualty of an overzealous biosecurity regime, and the Federation of American Scientists (FAS) devoted a web page to support him.

Even organizations that had historically been critical[40] of the proliferation of warfare research changed their tune. In the 1960s, the FAS, founded in 1945 by former Manhattan Project scientists concerned with the military's influence over science, had been one of the first to publicly advocate for arms control and for the United States to adopt a no-first-use policy for biological weapons. But in the post-9/11 era, FAS adopted the mantle of biosecurity: to educate scientists about the application of bioscience for warfare by terrorists. They even devised a web portal in 2006 called Case Studies in Dual Use Biological Research to encourage scientists to deliberate over how to balance national security and scientific freedom.

Undoubtedly the racialized bioterrorist imaginary that was embedded in the scaffolding of the biosecurity regime fostered this narrow social awareness among scientists—a focus on terrorist acquisition and "misuse" of presumedly benign research, rather than a broader political awareness of the use of biosciences in warfare. The post-9/11 narrative of victimhood may also have played a role, perhaps so successfully obscuring U.S. violence that even scientists working directly with dangerous pathogens failed to grasp the offensive implications of their work. Certainly, the "biodefense" rubric signaled a context quite different from the pre-ban era, with its prominent specter of U.S. deployment of biological weapons. The post-ban articulation of biological warfare research in the United States as merely defensive seems to have quashed the most obvious issues with the industry's relationship to warfare, thereafter muddling its continued destructive nature. Post-9/11 rhetorics of pre-emption and preparedness only further couched U.S. offensive action in the terms of defense from a possible terror attack on the United States.

It is perhaps understandable, if disappointing, that the record of U.S. scientists critical of the nation's post-9/11 biological warfare research industry was sparse. Even so, a few did speak out. Molecular biologist Barbara Hatch Rosenberg called out the U.S. biological program for treading the defensive-offensive line and criticized U.S. lack of compliance with the BWC. She published these points in a 2003 position paper titled "Secret Biodefense Activities Are Undermining the Norm against Biological Weapons" for the Federation of American Scientists Working Group on Biological Weapons. Prominent health practitioner-researchers Hillel W. Cohen, Robert M. Gould, and Victor W. Sidel published a piece in 2004 titled "The Pitfalls of Bioterrorism Preparedness: The Anthrax and Smallpox Experiences" in the *American Journal of Public Health*, highlighting the dangers of bioterrorism research and preparedness programs as well as the diversion of much-needed resources away from public health–focused research.

Howard Hughes Medical Institute researcher Richard Ebright brought attention back to biosafety.[41] In a 2006 *New York Times* article titled "Why Revive a Deadly Flu Virus?" he expressed disapproval of the research that re-created the 1918 pandemic flu strain due to the possibility of its release and a resulting pandemic: "I believe that this was research that should not have been performed. . . . If this virus was to be accidentally or intentionally released, it is virtually certain that there would be greater lethality than from influenza, and quite possible that the threat of pandemic that is in the news daily would become a reality" (quoted in Shreeve 2006).

These voices of dissent demonstrate that it remained possible to question hegemonic discourses of the bioterrorist Other and to promote prudence, rather than unconditional faith in technoscience and security regimes. They offer an accountability model for fellow scientists, an approach to one's research that considers its social impacts and proactively takes on that responsibility.

3
Co-opting Caregiving
Softening Militarism, Feminizing the Nation

The national Smallpox Vaccination Program announced on
December 13, 2002, was the result of an extraordinary policy
decision: to vaccinate people against a disease that does not exist
with a vaccine that poses some well-known risks. The rationale
for such a decision can be considered only against the backdrop
of the terrorist and bioterrorist attacks of 2001.
—Institute of Medicine (2005, 1)

We feel this smallpox program has been made for political reasons,
not public health reasons.... What it's doing is inflaming public fears.
—Lisabeth Jacobs of the California Nurses Association, as
quoted in Krupnick 2003

The U.S. national security apparatus enlisted not only the resources
of the biosciences, but also the vast disease surveillance and
response infrastructure of public health. To surveil for possible bioterror disease outbreaks, health institutions pulled data from
emergency room triage logs, 911 emergency calls, and pharmaceutical sales in local and state electronic health databases.
Health care practitioners underwent trainings to monitor patient
symptoms for indications of a bioterrorism event and to report

suspected bioterror outbreaks to law enforcement and intelligence agencies.

Many health scholars and scholar-practitioners have been critical of this encroachment of bioterrorism preparedness, which is known as the "securitization" of public health, seeing it as putting health resources in the service of national security aims, rather than public health aims. Prominent health practitioner-researchers Victor W. Sidel, Robert M. Gould, and Hillel W. Cohen (2002), in their article "Bioterrorism Preparedness: Cooptation of Public Health?," expressed skepticism about such "attempts to build long term public health capacity on the basis of what may well be exaggerated bioterrorism threats, while uncritically partnering with military, national security, and law enforcement agency-led preparedness strategies and programs" (82). Their article elaborated how this diverts attention from pressing endemic global health problems (e.g., TB, HIV, safe water supplies) and politicizes medical and public health decision making to conform with national security directives. Elsewhere I have detailed the negative health care outcomes of securitization, such as the view of patients as objects of suspicion rather than recipients requiring needed care (D'Arcangelis 2017).

A growing swell of health care workers also raised their concerns, particularly pertaining to what was perhaps the largest mobilization of public health for national security yet—the 2002 National Smallpox Vaccination Program (NSVP). The post-9/11 national security state had devised the NSVP as a bioterrorism preparedness/biodefense measure, slating over 500,000 military personnel serving in southwest Asia (where presumably they might be exposed to biological warfare on the battlefield) as well as 500,000 "frontline" health care workers (such as nurses and emergency response personnel) to receive vaccination against smallpox. And this was only the first phase of vaccination: to follow was the vaccination of additional health care workers and first responders (up to ten million), and, eventually, the entire U.S. public (General Accounting Office 2003).

The program constituted a massive response to nonexistent disease. As mentioned in the preceding chapter, smallpox had been eradicated in 1979; post-9/11 preoccupation with smallpox was the legacy of a post–Cold War specter that followed the disease's eradication.[1] But the 2002 program was also noteworthy because it enlisted civilians in a national security campaign. There was precedent for vaccination of the military against biological warfare threats—service members can be mandated to comply as part of their military duty (if they are active in regions of the world the U.S. military suspects has bioweapons capabilities).[2] The vaccination of civilians, however, was a new development. Their recruitment was thus voluntary; they could choose whether or not to participate.

In the beginning, most health workers expressed support for the program, believing that the vaccine would confer protection against smallpox in the event of an attack.[3] However, from the time the program was announced on December 13, 2002, to its start date of January 24, 2003, the NSVP lost support from vaccine administrators meant to administer the vaccine as well as health workers meant to receive it. Health workers' leading reasons for refusing vaccination were its side effects and belief that the risk of smallpox outbreak was not high enough (Wortley et al. 2006). Nurses, one of the primary groups targeted for vaccination, were at the forefront of the movement calling out not only the negative health consequences of the program, but also its links to post-9/11 U.S. imperial ambitions. They organized and published critiques in the months before the program's start date, directly lobbying government officials and speaking out via association newsletters and statements to the press.

Nurses' organizations known for their broader left/social justice views highlighted the program's connection with the U.S. military agenda to invade Iraq.[4] Over a dozen members of the Massachusetts Nurses Association penned a strong opposition statement to the NSVP in their January/February 2003 newsletter titled "Vaccinate against War, Not Smallpox." In it, they underlined the NSVP's role in war-mongering: "We say NO because vaccinating in the face of no known threat is wrong. It represents the use of health care as an extension of an aggressive military posture." Similarly, the California Nurses Association Board of Directors, in a position statement published on January 23, 2003, one day before the program's start date, contended that the program contributed to "efforts to generate support for an ill-conceived foreign military adventure [the invasion of Iraq]." Both statements argued that the vaccination program (and its vast mobilization of public health resources) was a tool of U.S. foreign policy.

Nurses' anti-imperialist critiques echoed broader left criticisms of the Iraq War, but also specifically exposed the health field's involvement in post-9/11 U.S. militarism. In doing so, they made an important contribution to anti-imperial critiques, which have largely overlooked the role science and health have played. I build on these nurses' insights about health and empire, focusing on the way the bioterrorism preparedness regime mobilized the caregiving valence of public health and images of feminized vulnerability to generate support for the NSVP. These gendered health narratives worked alongside familiar narratives of bioterror threat and patriotic duty to pressure health workers into enrolling in the program.

Conflating Health Care and Militarism

The backstory of the NSVP's formation reveals a history of tension over smallpox since the end of the Cold War. Public health typically prioritizes the management of existing disease, and therefore the field viewed smallpox as a lesser priority once it was eradicated. The national security field, on the other hand, viewed smallpox as a military threat whose successful eradication only increased the vulnerability of the national population—lack of exposure to smallpox meant that the nation's immunity to the disease had declined (although the degree of loss was unclear, since the data on smallpox was outdated and biomedical circles could not reach consensus on the length of immunity the vaccine conferred).

Concerns about susceptibility to smallpox had escalated under Clinton-era civilian biodefense, culminating in the highly influential Dark Winter exercise in the summer of 2001, which comprised one of several mock terrorist scenarios policy makers conducted at the turn of the twenty-first century.[5] Dark Winter simulated a smallpox attack, based on relatively high transmission rates and susceptibility of the population to smallpox—the scenario used an immunity value of no more than five years, at which point immunity waned significantly.[6] The scenario presented a stark picture of devastation: within ten weeks of exposure to smallpox, three million would be infected and about one million dead (O'Toole, Michael, and Inglesby 2002).[7] Many health researchers contested this "worst-case scenario" and discouraged basing policy on these numbers.[8] Nevertheless, it greatly influenced ideas of smallpox preparedness, particularly in national security circles, and created momentum for vaccine production and vaccination in anticipation of an outbreak.

The post-9/11 push for precautionary smallpox vaccination—embodied by the NSVP—was the culmination of this militarized approach to smallpox. The NSVP's target vaccination numbers reflected the influence of the alarmist worst-case scenarios pushed by defense pundits (namely, Vice President Dick Cheney and several Department of Defense officials); these numbers were much higher than those proposed by the Centers for Disease Control and Prevention (CDC), despite the fact that the latter was in charge of carrying out the actual vaccinations (CDC 2002; Cohen and Enserink 2002). The military had not only triumphed over the smallpox agenda but enlisted public health to implement its vision.

When Bush announced the program on December 13, 2002, he emphasized its health dimension as much as its security one: "We will continue taking every essential step to guard against the *threats to our nation* and I deeply appreciate the good efforts of state and local health officials who are facing difficult challenges with great skill. The actions we are taking together will help safeguard the *health of our people* in a measured and responsible way" (Bush 2002b; emphasis added).

In articulating this effort as having a positive outcome for the health field—that is, the promotion of the "health of our people," Bush turned the national security apparatus's mobilization of the health field into a dual benefit—for the nation's health as well as security. Legal scholars David P. Fidler and Lawrence O. Gostin (2007) call this placement of disease alongside national security the "synergy thesis";[9] it yielded new surveillance initiatives oriented toward improving early detection of both bioterrorist attacks and infectious disease outbreaks (for example, the National Biosurveillance Initiative of 2004). This synergy orientation would eventually integrate health and security among a variety of diverse hazards—an "all-hazards" approach. The Pandemic and All-Hazards Preparedness Act of 2006 exemplified this dramatic convergence of domains. Its stated purpose was "to improve the Nation's public health and medical preparedness and response capabilities for emergencies, whether deliberate, accidental, or natural"—with "deliberate" signifying bioterrorism, "accidental" referring to lab mistakes, and "natural" meaning naturally arising outbreaks. The NSVP represented an important early site where the Bush administration attempted to merge these disparate domains.[10]

A week into the NSVP's start date, Dr. Anthony S. Fauci, director of the National Institute of Allergy and Infectious Diseases, articulated the thesis that public health and biodefense could join forces in his 2003 testimony during the *Smallpox Vaccination Plan: Challenges and Next Steps* hearing: "At the end of the day we believe this [research on biological agents such as smallpox to produce vaccines and other countermeasures] will have two major accomplishments. One will be that it would effectively defend us against the microbes of bioterror. But also, since bioterror agents are really emerging and reemerging diseases that resemble very much the naturally occurring diseases, so that what we learn for biodefense will have important implications for decades and decades to come in our approach toward emerging and reemerging diseases."

Fauci's statement, echoing the Bush administration's rhetoric of converging biodefense and disease management interests, belied

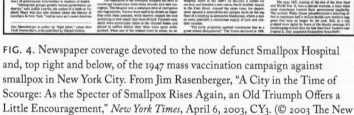
FIG. 4. Newspaper coverage devoted to the now defunct Smallpox Hospital and, top right and below, of the 1947 mass vaccination campaign against smallpox in New York City. From Jim Rasenberger, "A City in the Time of Scourge: As the Specter of Smallpox Rises Again, an Old Triumph Offers a Little Encouragement," *New York Times*, April 6, 2003, CY3. (© 2003 The New York Times Company. All rights reserved. Used under license.)

the reality—the de facto subsumption of public health by the national security apparatus.

The news media contributed to this visage of synergy by circulating archived photos of mass vaccination campaigns from the 1940s. The news article "A City in the Time of Scourge: As the Specter of Smallpox Rises Again, an Old Triumph Offers a Little Encouragement" (Rasenberger 2003), which appeared in the *New York Times*, included two photos depicting a 1947 vaccination campaign in New York City (see figure 4, bottom left; right). It also included a third photo (top left) depicting the long unused, dilapidated Smallpox Hospital—the caption identified its origins in 1856.

The two historical photos of mass vaccination in 1947 recalled a period of rampant smallpox outbreaks when these campaigns were an important tool of public health. The 1947 mass vaccination campaign in New York City was the last the United States conducted, as there would be no more smallpox outbreaks after 1949. Routine vaccination of civilians (as infants) had ended by the early 1970s,[11] thirty years before the NSVP. The author's revival and placement of old photos in a 2003 article about smallpox—a context driven primarily by the goals of biodefense, without the presence of the disease *anywhere*—potentially blurred for readers the distinction between vaccination in the context of a present health danger and vaccination as a form of bioterrorism preparedness.

Photographs can seem to purely reproduce reality (that a threat exists and that vaccination is advised), easily leading viewers to miss the fact that journalists exercise choice in selecting which images to include and where to include them. Photographs included in news coverage, moreover, can convey meaning in ways that text cannot—in particular eliciting strong emotional reactions from readers (in this case, likely fear but also safety).[12] The invocation of the health connotations of earlier mass vaccination campaigns, I argue, re-narrated post-9/11 smallpox preparedness and the NSVP as a continuation of the history of public health campaigns against smallpox, lending the post-9/11 programs a caregiving valence.

In addition to photograph selection and placement, their repetition can further enhance the effects of the image's content (Corrigall-Brown 2011). Journalists covering smallpox from late 2002 to early 2003, the height of the NSVP, frequently incorporated the historical smallpox photos—in different articles across media outlets.[13] In doing so, I suggest, they provided extensive cover for and distraction from the NSVP's militaristic origins and aims.

As I briefly described in the introduction, U.S. public health practice has long been informed by a national imaginary that views communities of color (as well as other groups marginalized by U.S. society) as diseased Others that threaten national health. The ascendance of germ theory in the late nineteenth century had marked bodies of color, already seen as biologically inferior to and different from the white body, as infectious disease vectors.[14] As a result, public health authorities focused their efforts on protecting the white national body from them. U.S. public health officials treated blacks, constructed as having an innately heightened sexual appetite, as more prone to spreading venereal disease than whites (Jones 1993). Mexican American and Asian American immigrants were also associated with disease, which was attributed to their "backward" and "less civilized" cultures, and thus they were constructed as sources of deadly contagious diseases such as typhoid, cholera, and plague (Molina 2006). This led, quite frequently, to disease control campaigns that subjected these populations to harsh disease control regimens—hygiene and sanitation measures, forced immunization and drug treatments, as well as quarantines and other forms of spatial control (Kraut 1994; Stern 1999).[15] I and others have demonstrated more recent manifestations of racialized disease Othering, with respect to Chinese Americans during the SARS outbreak of 2003, for example.[16]

The figure of the racialized Other as public health threat found purchase in discourses underpinning biodefense. Cultural studies scholar Neel Ahuja (2016) has described the circulation of horror-inducing images of smallpox victims during the Dark Winter mock terrorist exercise at the turn of the century. The images were primarily taken in India, Bangladesh, and the African continent in the 1970s, and Ahuja notes that the original purpose of the photos was for local health efforts (aiding the public to recognize cases), but became, in the context of Dark Winter, appropriated to "communicate a racialized, border-crossing contagion that proliferates within the body as days progress" (163). Ahuja argues that these images invoke tropes of racial threat—namely, the inscrutable Asian,[17] and that the trope's affective power served to generate fear and uncertainty and thereby justify state action with respect to biodefense.

Images of diseased bodies of color appeared in news coverage of the NSVP,[18] and journalists also elaborated this figure in text. "A City in the Time of Scourge" emphasized that successful vaccination campaigns hinged on "evenhanded" health authorities as well as public trust in the government, and argued that this lack of trust was a primary reason for the faltering NSVP. In discussing public trust, the article outlined the history of smallpox vaccination, going back to the 1800s, and focused on immigrants: "Thousand[s] of malnourished immigrants arrived daily, cramming into tenements. Epidemics of infectious disease swept through the city with lethal efficiency" (Rasenberger 2003).

The journalist, by following the sentence about large numbers of immigrants with a sentence about epidemics, implies a causal relationship—that immigrants spread disease. Further, the reference to tenements invokes the historically racialized specter of impoverished immigrant hordes arriving and overtaking U.S. cities. The Irish population (internally colonized by England) comprised a large portion of immigrants in the first half of the nineteenth century, followed by large numbers of Chinese and peoples from central, eastern, and southern Europe. All of these populations were negatively racialized at the time—albeit to different degrees, and stigmatized accordingly.[19] Their construction as disease vectors was one part of this larger system of marginalization.

The article went on to express a common criticism of immigrant interaction with public health authorities, quoting historian Judith Leavitt that "immigrant groups greatly feared government authority" and that "as soon as it came down too strongly, especially in the guise of uninformed health officials, they would resist. They'd hide cases, and the epidemic got worse" (Rasenberger 2003). In characterizing immigrant behavior as driven by "fear" and "resistance," the article further placed responsibility on them for spreading disease. It suggested that immigrants were not just unwitting public health threats but acting deliberately in their refusal to comply with public health measures. The news article failed to contextualize immigrants' fears against the backdrop of U.S. public health authorities' harsh treatment toward marginalized groups in this period.

Ironically, Leavitt herself has highlighted the social inequities that typically characterized public health campaigns; she is best known for her 1996 book on Mary Mallon (a.k.a. "Typhoid Mary"), which illustrates the harsh treatment Mallon received as a working-class female Irish immigrant at the turn of the twentieth century.[20] Leavitt describes the anti-immigrant sentiment that was "widespread among health officials" (117) and the connection to larger patterns of U.S. social hierarchy that subject marginalized women in particular to the "stings of discrimination" (165). Leavitt thus centers the structural inequities perpetrated by U.S. public health as crucial to understanding the reactions of marginalized groups to these authorities.

Leavitt's approach to the topic lies in direct contrast with the way she is quoted in the article, suggesting that the journalist quoted her out of context. Even when the news article discussed the sociopolitical basis of immigrant fears, it attributed them to "uninformed health officials" and corrupt government officials, for example highlighting medical workers at the infamous Smallpox Hospital: "Nurses were said to be drunk on liquor they stole from patients, while doctors treated the sick according to how much they were bribed" (Rasenberger 2003). This scapegoating of individual health workers and hospital sites minimized what was actually systematic mistreatment of these immigrant groups. Thus, the article maintained the reputation of U.S. health authorities, while black-boxing racialized immigrant populations as obstacles to the public's health.

The article's invocation of a noncompliant racialized Other as public health threat set the stage for its opposite—the compliant subject. After its reproduction of a racist, xenophobic view of immigrants as disease-spreading Others, the article devoted much of its space to extolling the 1947 vaccination, attributing the campaign's success in large part to public compliance: "New Yorkers lined up for blocks in front of dozens of vaccination clinics, while thousands of volunteers took turns giving shots. Within two weeks, five million people had been vaccinated; after a month, the number had risen to 6,350,000." This set up a comparison between past public compliance and the resistance that met the post-9/11 vaccination program: "The current White House administration's faltering effort to vaccinate half a million health care workers suggests that [public trust in the government] may no longer be the case. Still, in a city gripped once again by fears of the deadly scourge, it's encouraging to note that the last time New Yorkers confronted it, they acquitted themselves beautifully" (Rasenberger 2003). These were the last sentences of the article, and summarized the point the more than 1,300-word trip down memory lane had been building to: vaccine compliance.

The article appeared at the precise moment that the legitimacy of the NSVP was most threatened. As mentioned at the start of the chapter, many health workers, in particular nurses, were reluctant to get vaccinated against smallpox because of the program's tie to alarmist discourse about Iraqi threat, and also because of the vaccine's side effects. By the time the article was published in early April, significant adverse events had emerged among vaccine recipients, ranging from minimal to life-threatening.[21] These adverse events garnered widespread press attention and generated increased resistance to the program. While most view adverse reactions to vaccines as tolerable when vaccines confer protection against the ravages of disease, with respect to smallpox, a *nonexistent* disease, these adverse reactions constitute unnecessary harm. When nurses contested the vaccine for smallpox, they did so because the disease was no longer present to cause harm, not because they opposed vaccines in general (a distinct view belonging to the highly publicized anti-vaccination movement).[22]

"A City in the Time of Scourge" recapitulated a nostalgic narrative of compliance to suggest that noncompliance was the most significant hindrance to public health campaigns. In doing so it arguably reframed health workers' substantive criticisms of the

FIG. 5. This 1947 photograph circulated in many 2002–2003 newspaper articles on smallpox. The photograph's original description read: "Dr. Walter X. Lehmann, left, and Dr. Kurt L. Brunsfeld, right, vaccinate two unidentified women for smallpox, April 14, 1947, as others await their turn in New York City Health Department building." (File 470414023, AP Photo/Tony Camerano.)

NSVP (its politics, vaccine hazards, and the lack of smallpox threat) in the disparaging terms of the narrative it had laid out— as the product of ignorance, recalcitrance, and overreaction.

Illusions of Vulnerability: Mobilizing White Women and Children

The linguistic construction of the compliant subject in "A City in the Time of Scourge" (Rasenberger 2003) was mirrored in the photo on the right (an enlarged version of which I reproduce in figure 5).

The photo (taken by the Associated Press) connoted an image of compliance through its depiction of white[23] women pushed up against each other in line, smiling, posing, and showing off their vaccinated arms. Unlike the recalcitrance of the diseased Other, the white women lined up for vaccination appear to be positively inclined toward it. The academic literature on public health has demonstrated the way that white populations have historically been the intended beneficiaries of U.S. public health—constructed as innocent populations vulnerable to threats from Others, be that immigrants, people of color, or other marginalized groups who are portrayed as dangerous disease vectors. White women, moreover, have typically been cast as caretakers responsible for ensuring the health of white populations—whether that be as mothers and wives or in formal occupations as nurses (Leavitt 1996; Shah 1999).[24] White women, then, epitomize white vulnerability but also embody the tool for further safeguarding whiteness. Given this context, it is not surprising that such a photo exists of white women seeking inoculation to smallpox in the 1940s—public health programs largely served them and their presumably white families. But the journalist's reproduction of the photo in the post-9/11 era recast it within a new landscape of meaning.

The Bush administration had cultivated gendered narratives of the United States as victim. A prominent story line framed the September 11 attacks through the metaphor of sexual violation—the United States as a feminized (white) nation violated by the brown (male) terrorist Other (A. Cole 2007). This frequently interlocked with another common figuring—the need for a masculinist state to protect its feminized citizenry, as Puar and Rai (2002) have articulated in describing the war on terror as a project seeking to rectify the "emasculation of the white male state (signified by the castration of the trade towers on 11 September)" (138). In many of these nationalist imaginings, white women represented the ultimate symbol of white fragility in need of protection from various racialized threats.[25]

The construction of a feminized U.S. vulnerability in reference to 9/11 served to disguise and justify U.S. aggression abroad and other imperialist endeavors. Postcolonial gender studies scholar Gil Hochberg (2015) has described the way that nationalist narratives deploy tropes of femininity to conceal state violence, specifically describing how the Israeli army has used images of its female soldiers to "hide militarization behind a mask of feminine tropes" (12). A similar effect occurred, I contend, when U.S. news media mobilized the image of white female vaccine recipients in articles about the NSVP: it furnished a (white) feminine face to the program, softening its militaristic valence and validating the government's framing of the program as defensive, rather than linked to the aggression of the war on terror. In this way, the media image of white women seeking the vaccine's protective function reinforced the foundational rationale of the NSVP—U.S. vulnerability to smallpox and bioterrorism more generally.[26]

The photo also appeared in a later article by Richard Perez-Pena that defended the program against its critics: "Checking City's Archives to Solve a Medical Mystery," published in the *New York Times* on October 3, 2003. The article took on the controversy that had threatened to undermine the program—vaccine adverse events, including possible vaccine-related deaths. This article, which emerged months after the news first broke, focused on the three vaccine recipients who had died from myocardial infarctions (heart attacks). It discussed the investigations of New York City officials who had gone back to the 1947 vaccination campaign records and found no incidence of a rise in heart attacks.

This finding was meant to demonstrate that the cardiac complications were not attributable to the vaccine. However, other studies had shown evidence of a link to the vaccine. Further, some infectious disease specialists believed that cardiac events may have been missed in the earlier records or were *new* complications of the vaccine (and thus would not have appeared in the earlier records).[27] There was, therefore, a larger debate about the causal connection between the vaccine and the cardiac-related deaths—a debate the journalist failed to mention. Instead, he highlighted findings that exonerated the vaccine, and argued for how safe this vaccination campaign was over previous ones (Perez-Pena 2003).

The text's unambiguous support for the program was punctuated by the messaging of the photo that stood alone in the center of the article, dominating the messaging. It was, moreover, a cropped version of the original Associated Press photo (see figure 5)—the left half. The portion shown in this article accentuated the narrative of feminized vulnerability, depicting exclusively white women (except for two young white boys presumably with their mother) receiving vaccination from a white male doctor—the cropping had cut out all four female nurses and the few male vaccine recipients who were previously visible in the right side of the uncropped photo. This cropped photo, in leaving intact only white male medical authorities and female vaccine recipients, invoked the authoritative relationship between the white male doctor and the compliant white female patient—an image of white fragility and concomitant protection that, at a critical juncture, reinvested the program with the connotation of provisioning safety.

This photo of white women seeking vaccination appeared in other articles, at various points during the debates on smallpox (before, during, and after the implementation of the vaccination program). Its repeated use, and that of additional photos of white (or mostly white) women getting vaccinated, collectively constructed a trenchant image of national vulnerability. Like other cultural studies scholars, I take seriously the role of mass culture in extending state ideologies and garnering support for them (Althusser 1971). While in some cases the state exerts direct control over the news media (for example, in the case of the media blackout during the 2003 invasion of Iraq), in other cases the control is less overt (or intentional), and comes in the form of the subtle constraints that state-generated discourse exerts on how journalists think and write. I do not know whether the government dictated the use of these historical smallpox images. I suspect that the photos were either provided to journalists by government officials or were the ones journalists could most easily obtain from photo archives. Regardless of the specific way the photos made their way into the mainstream news media, I have focused on their effects, and argued that they convey particular themes and meanings that more often than not lent support to the

FIG. 6. "Children receiving smallpox vaccinations in Virginia in 1946. A new vaccination campaign has been announced." From William J. Broad, "Bush Signals He Thinks Possibility of Smallpox Attack Is Rising," *New York Times*, December 14, 2002, A13. (Courtesy National Archives photo no. 245-MS-1168L.)

government's rationale for vaccination—painting a picture of a United States in urgent need of preparation for the ominous and unpredictable threat of bioterrorism.

The image shown in figure 6 was prominently centered in a *New York Times* article titled "Bush Signals He Thinks Possibility of Smallpox Attack Is Rising" (Broad 2002), published on December 14, 2002, which was the day after President Bush announced the NSVP. The photo dates from 1946, and depicts white[28] children receiving the vaccine during a vaccination campaign in August 1946 in Virginia. The text advocates strongly for the government's vaccination program, outlining its rationale and defending it against its critics.

Children, particularly girls, are understood as even more vulnerable and in need of protection than adult women—what better image to invoke the protective connotation of vaccination? Feminist political scientist Cynthia Enloe (2014) has highlighted the patriarchal protectionism embedded in the trope of "women and children": it represents the most innocent and vulnerable members of a nation, who require refuge from the barbarism of the threatening Other who would target these innocents. Thus, the symbolic deployment of young white girls[29] here conveyed the NSVP as providing health and well-being, punctuated, as before, by the authoritative connotations of white medical authorities delivering the vaccine.

Counternarratives: Reframing Vulnerability and Reclaiming Health

In some ways, this vulnerability narrative, and the complementary narrative of the bioterrorist Other, failed. Even before the NSVP began, nurses' organizations and other health care institutions expressed skepticism toward the Bush administration's alarmist rhetoric of an Iraqi smallpox threat, suggesting that the NSVP served no real protective function. Once the program was implemented, the growing number of adverse reactions to the vaccine further unmasked the protectionist rationale of the NSVP.

Recall that the Bush administration and health officials presented the NSVP, and biodefense in general, as serving the dual benefit of security and health. But critics placed front and center the issue of vaccine hazards, which endangered the health workers receiving the vaccine as well as their patients, who would be exposed for up to twenty-one days through their inoculated caregivers.[30] This made the government's claim that the NSVP provided safety difficult to maintain, particularly when juxtaposed against the purely hypothetical risk of a smallpox attack. The California Nurses Association was especially vocal in conveying this critique in a January 23, 2003, position statement opposing the NSVP: "There is no proven evidence of the likelihood of a smallpox attack. . . . However there are known dangers from a massive smallpox vaccination program both to caregivers and their patients. Those include severe life-threatening skin reactions, brain inflammations, and non-life threatening reactions to the vaccination."

This statement demonstrates the way that nurses' organizations flipped the program's premise of security on its head, stressing instead the vulnerabilities it produced.

These counternarratives made headway in the national security debate and, with the aid of media exposure, undermined the government's narrative. In addition, nurses' successful organizing and refusal to participate in the program contributed to its disintegration.[31] The tally reflected the program's drastic decline: of the 500,000 frontline health care workers slated to be vaccinated in the first thirty days, only 7,543 had been (less than 2 percent), and in ten weeks only 31,297 (about 6 percent) (Government Accounting Office 2003). In May 2003 the Advisory Committee on Immunization Practices (ACIP), which advises the government on immunization policy, recommended a pause in the program to assess safety; in June it recommended the program cease altogether. Even though the program did not heed this recommendation, few health workers continued to volunteer for vaccination. Weekly vaccination numbers dwindled to a handful by July 2003 (Institute of Medicine 2005). The program abandoned its planned second and third phases.

The Institute of Medicine stressed the impact health worker critiques had made on perception of the program: "Mass media reports showed a downward shift in public perception about the level of risk of smallpox release and therefore a decreased motivation to receive the vaccine" (Institute of Medicine 2005, 51). An official government report, the GAO report to the Senate Committee on Governmental Affairs in April 2003, had also highlighted the impact of these critiques, suggesting one of the two major hindrances to the program was the "hesitation on the part of the two main groups needed to participate in the program— the state and local public health authorities and hospitals needed to implement it, and the health workers needed to volunteer to be vaccinated" (Government Accounting Office 2003).[32] Thus, just as the dominant rhetorics of text and image I analyzed in this chapter promoted the vaccination program, so too did the rhetorics employed by the program's critics embolden resistance and refusal.

My analysis has highlighted the tactics of the U.S. national security apparatus in mobilizing not only a caregiving valence, but also
narratives of feminized vulnerability. My analysis has also highlighted the role of news media in reinforcing the protective connotations of the NSVP and amplifying the sensibility of national
vulnerability in its coverage of smallpox, in particular through
images of white women (and girls) as eager vaccine recipients from
an earlier era of mass vaccination. But the gendered dimensions
of the NSVP were not only on the level of rhetoric—they also had
a material dimension.

The primary target for vaccination—that is, frontline health
care workers—constituted a feminized field comprising predominantly women.[33] On the one hand it is important to note the unfortunate irony that the symbolic recruitment of white women into
an idyllic picture of national security belied the reality of their
material incorporation into a program rife with risks. But, on the
other hand, it is also important to note that this symbolism did
not reflect the actual racial composition of the frontline health field
enlisted by the NSVP.

Women of color and working-class women are disproportionately concentrated in frontline work,[34] which involves the most
direct patient care and the most exposure to the dangers of triage
(assessing and treating patients with potentially unknown afflictions, in decidedly uncertain situations, before directing them to
appropriate services and doctors). It is less prestigious and lower
paid, and reflects the race and classed patterns of care work
(D'Antonio 2010; Melosh 1982; Reverby 1987). The NSVP, then,
further burdened groups of women already exposed to the most
dangerous aspects of health care work. The rhetorical apparatus
upholding the NSVP echoed this marginalization—it reproduced
white femininity as standing in for national vulnerability, and
therein left out women of color and working-class women from
this (albeit problematic) protectionist narrative, affording them not

even symbolic protection. Simply put, the trope of feminized vulnerability is an utterly whitewashed one.

That the dominant narrative belied the harms the NSVP meted out to white women as well as more marginalized women is sadly unsurprising, given the at times high-flying feats of dominant national security discourse. The war the NSVP was connected to—an unprovoked invasion of Iraq that cost the lives of countless Iraqis[35] as well as U.S. military personnel[36]—was couched as "preemptive" and "defensive." At the time of the NSVP, the U.S. national security apparatus had produced so elaborate a narrative of Iraq as a threat and the United States as vulnerable that it turned victims into oppressors and vice versa. The painting of a smallpox threat and need for vaccination, then, easily fit into this imperialist-nationalist U.S. narrative.

The problem with the NSVP, then, derived from its connection to violent imperialism and not from the mobilization of health for security per se. The overlapping of health and military domains has occurred in many iterations throughout history, as much in revolutionary anti-imperial health campaigns such as the free health clinics of militarized social justice groups like the Black Panthers[37] as in unjust programs such as the military medicine units attached to U.S. and European colonial medicine endeavors that provided care to colonists and the local labor force so as to ensure the success of the colonial project.[38] The relationship between military and health domains is thus secondary to their relationship to power—a just program is one tied to liberation, an unjust one to imperialism.

The solution, I submit, lies in imagining alternatives to imperialist military-health configurations. Recall that the critical health practitioner-scholars I mentioned in the first section of this chapter critiqued the imperial and transnational dimensions of U.S. biodefense. Their critique of the militarization of health revolved specifically around the way in which it negatively impacted global health equity—by diverting resources from pressing endemic global health problems. They were concerned with the way the U.S. national security apparatus pulled focus from global health.

Nurses' organizations also took on a global focus in their critiques of the NSVP—they were concerned not solely with the risks of the smallpox vaccine to the U.S. national population, but also with the global implications of supporting a biodefense program that would fuel an unjust war against Iraq. The Massachusetts Nurses Association stated: "We must use our health care abilities to build an international commitment to peace and human rights. Let the example of smallpox eradication be used to further cooperation. . . . We have pledged first to do no harm. . . . We will accept the smallpox vaccination when it is part of a worldwide effort to eradicate the disease. In that event the health care workers of Iraq would be inoculated as well" ("Vaccinate against War, Not Smallpox" 2003).

This statement centered health care as a means to build international relations founded on peace and human rights, rather than militarism and oppression. It suggested an alternative to biodefense, rebuking the narrative of Iraq as a threat whose citizens deserve no protection as well as the complementary notion that the U.S. population is uniquely deserving of health and security. The nurses' statement proposed that Iraqi health and lives be accorded utmost consideration. In doing so, the statement opposed bio-imperialism in the most fundamental sense—it called out the power dynamics between the United States and Iraq, advocating that biological resources, namely, the vaccine (in theory), not be hoarded by the United States, but should also benefit Iraqis. This rejection of U.S. hegemony over biological resources paved the way for health and security systems aimed at global equity.

Conclusion: Ongoing Challenges

The security field's mobilization of public health was not nearly as seamless as its harnessing of the bioscience field that was the focus of the preceding chapter. Vocal critics from the health field threw a wrench into the onslaught of the bio-imperial biodefense apparatus and its narratives of threat and vulnerability. Their achievement was extraordinary, particularly during a period of extreme patriotism, when it was difficult to wage any criticism of the U.S. national security apparatus whatsoever. On the other hand, as important as this interruption was, the larger bioterrorism preparedness agenda in many ways succeeded.

Despite its limited success, the NSVP further entrenched a military-health nexus that not only recruited public health into the U.S. state's bio-imperial apparatus, but also framed caregiving in the distorted terms of post-9/11 national defense. By the time of the program's unceremonious demise, the CDC and the Department of Health and Human Services of which it was a part showed little trace of their earlier reservations about precautionary vaccination and the high numbers targeted by the NSVP. The statements released at the close of the program by health officials instead articulated the program's success—in, moreover, the terms of post-9/11 preparedness. For example, U.S. Health and Human Services Secretary Tommy Thompson sanguinely announced on January 29, 2004, that the "vast majority" of states were now prepared to immunize all their residents in ten days if there were a smallpox bioterrorism attack (paraphrased in a *Nuclear Threat Initiative* news article; see McGlinchey 2004). By framing the goal in terms of preparedness and skirting the issue of the vaccine risks and health hazards created by the program, Thompson reflected the solidification of the preemptive security focus in public health.

Moreover, the seeming reluctance of the CDC and top public health officials to address the flaws that emerged throughout the program's implementation reveals a disconcerting acquiescence to the post-9/11 national security apparatus and its terms. After the call by the ACIP in June 2003 to cease vaccination, CDC director Julie Gerberding reiterated the agency's commitment to proceeding with smallpox vaccination. She repeated dominant national security rhetoric and maintained the rationale for the program: "We have no information since December, that would suggest that the threat of a smallpox attack is any less now than it was last year, and our President made a policy decision based on that information, that we needed to be able to protect our country should we have a smallpox attack" (CDC 2003b). It is quite astonishing that despite continued lack of evidence of threat from Iraq, and in the face of the many criticisms of the NSVP from both security and health perspectives, the director relinquished the opportunity to reject the mandate of biodefense.

Thus, in the end, the NSVP, despite its obvious failures, was quite successful in consolidating a military-health nexus that recruited caregiving—in both its rhetorical and material dimensions—to biodefense. Moreover, the government's and the news media's gestures to feminized white vulnerability, even if largely unable to legitimize the NSVP as a program of care and protection, arguably furthered the picture of U.S. vulnerability so central to the larger U.S. national security agenda. In contrast with the well-worn post-9/11 narrative of bioterror threat, these allusions to national vulnerability drew upon many layers of surreptitious suggestion, and their continued role in this long-term agenda requires continued vigilance. In particular, the larger bioterrorism preparedness agenda eventually went global, hijacking the discourse and institutions of global health along the way, which is the subject of the next chapter.

4

Preparedness Migrates

Pandemics, Germ Extraction, and "Global Health Security"

SARS has clearly shown how inadequate surveillance and response
capacity in a single country can endanger the public health security
of national populations and in the rest of the world.
—CDC and WHO epidemiologist David Heymann, "The
International Response to the Outbreak of SARS" (2004, 1128)

Is plugging into the global knowledge network the only scale of
possible engagement for securing human security, or can an
exception be made for the political legitimacy of collective life and
its ethical status in the tropics?
—Sociocultural anthropologist Aihwa Ong,
"Scales of Exception" (2008, 125)

On March 12, 2003, the World Health Organization (WHO) issued a global alert for severe "atypical pneumonia." Three days later, the WHO designated the new virus "severe acute respiratory syndrome" (SARS), declaring it "a worldwide health threat" and issuing a rare air travel advisory (WHO 2003d). The first known cases of SARS had appeared in Guangdong Province, China, in November 2002. By late February 2003, cases of the new disease had been found in neighboring Vietnam and Hong Kong and as far away as Canada and Ireland. The possibility that SARS's pathogenicity (ability to cause disease) and transmissibility between humans might make it the next pandemic (an infectious disease spreading widely across the globe) garnered international concern.

In the United States, the first cases of SARS were reported on March 20, 2003. On April 4, President Bush issued an executive order adding SARS to the list of quarantinable diseases, giving the Centers for Disease Control and Prevention (CDC) power to detain and examine persons suspected of carrying SARS. The CDC triaged ill passengers at airports and worked with state and local governments to alert travelers to risks and prepared health care delivery systems to recognize SARS (Institute of Medicine 2004). By April 17, upwards of 200 suspected cases of SARS were reported to the CDC, upon further investigation dwindling to less than 40 (CDC 2003d).

On July 5, the WHO announced the end of the epidemic—in no small part because SARS turned out to have low rates of transmissibility, morbidity, and mortality. In total it had killed 774 people between late 2002 and mid-2003 across two dozen countries in North America, South America, Europe, and Asia (WHO 2003c). While SARS death tolls did not reach as high as public health officials had originally feared (in the United States there were no SARS-related deaths at all), relief was short-lived when a deadly strain of influenza emerged—H5N1.

The H5N1 avian flu strain, or "bird flu," as it was often called, was first found in a domesticated goose in Guangdong Province, China, in 1996. Its first documented infection of humans occurred in Hong Kong in 1997. The Hong Kong government had responded swiftly, killing millions of domestic birds (on poultry farms and in markets) and eliminating the virus's primary host. The outbreak ceased by the end of the year, having caused six human deaths (out of eighteen total cases). Six years later, H5N1 re-emerged in the same region, and by December 2003 it appeared in Vietnam and Thailand, killing over thirty people in 2004 (WHO n.d., 2012).

Unlike SARS, influenza was not a new disease. However, its mutability and adaptability have resulted in many new strains each year—infecting a variety of hosts, such as humans, birds, and pigs. Influenza strains that affect humans lead to an estimated hundreds of thousands of deaths annually. On occasion, flu strains develop that possess high pathogenicity and transmission rates in humans—causing a pandemic. Pandemic strains emerge when a nonhuman strain (often avian or swine) acquires the ability to infect humans, who have little or no immunity against it.[1] Historically, flu pandemics have caused deaths in the millions worldwide.[2]

H5N1's spread from southern China to Southeast Asia, and its high pathogenicity in humans, suggested that it might transform into a human flu pandemic. Like any future catastrophic scenario, a flu pandemic is difficult to predict. Health officials around the world vigorously debated the likelihood of an H5N1 pandemic: those who believed pandemic transformation was a likely possibility pointed to the increasing numbers of human infections with avian influenza viruses since the mid-1990s;[3] others argued that its lack of contagiousness mitigated the concern.[4]

U.S. public health authorities, in step with the preemptive mode of the war on terror, pushed forward with pandemic flu preparedness plans that drew on worst-case scenario models projecting the disease's rampant spread and millions of deaths across the nation.[5] Four federal flu plans in 2005–2006[6] outlined a far-reaching approach that spanned the domestic to the international, including augmenting vaccine stockpiles and enhancing international disease surveillance capacity, as well as more contested measures such as large-scale geographic quarantine, border screening, and greater military role in compliance enforcement.[7]

In the end, H5N1 did not become the pandemic imagined—the death toll numbered 250 worldwide between 2003 and 2008,[8] waning by the time a new strain (H1N1 "swine" flu) emerged in 2009 and took center stage. The United States saw no H5N1 outbreaks in either birds or humans during that period (CDC n.d.d),[9] but the scare would set in motion an enduring pandemic preparedness infrastructure—reorienting post-9/11 public health from anticipation of a bioweapons attack to the more common form of disease threat, namely, natural disease emergence. Pandemic preparedness spurred the United States to improve global management of outbreaks in ways that, I show, prioritized U.S. and Global North health interests to the detriment of Global South regions. This chapter examines the U.S. and global narratives that buttressed this bio-imperial system, namely, U.S. exceptionalism and global health security. It also focuses on how the Global South, in particular China and Indonesia, challenged this model of U.S. and Global North control, exposing its cracks and carving out new avenues for equitable global disease governance.

Precursors to Pandemic Preparedness: Eurocentrism and SARS

Post-9/11 concern over disease pandemics was part of a longer history in public health—the culmination of the "emerging diseases worldview" that came to the fore in U.S. public health in the late 1980s.[10] This view posited new and formerly dormant infectious diseases—that is, "emerging infectious diseases"—as a rising, serious health threat. Proponents of this worldview argued that although there had been a brief respite from infectious disease in the United States in the mid-twentieth century, a dramatic resurgence was under way: increased incidence and geographic spread of diseases long thought to have dwindled (such as malaria); new variants of old diseases such as multi-drug resistant tuberculosis; and even new diseases such as Ebola and HIV/AIDS. They pointed to a wide range of factors—from public health deterioration and microbial evolution to environmental change and food system vulnerabilities.[11] But they found especially troubling these diseases' ability to cross borders.

By the 1990s, the "emerging diseases worldview" came to dominate infectious disease control approaches in the United States, as well as at the World Health Organization—elevating the urgency of global health, that is, health problems that transcend national boundaries and necessitate global cooperation.[12] In the decade that followed, "global health security" framed the cross-border spread of infectious diseases as a type of security threat, and further prioritized global health over national interests and sovereign authority (Brown, Cueto, and Fee 2006; Weir 2014).

Global health security and its predecessor, the emerging diseases worldview, both mobilized a fundamentally Eurocentric framework that assumed that the Global South was to blame for the rise of new and resurgent infectious diseases.[13] This premise was the legacy of colonial health paradigms originating in western Europe in the late nineteenth century (e.g., tropical medicine) that primarily focused on the colonies as the cauldron of exotic diseases threatening the colonial enterprise (the health of the colonizers and the colonized labor force) (N. King 2002, 772–773; Levich 2015, 714–715). Accordingly, the global health security regime focused on the diseases and health concerns of interest to the West and other Global North nations.

When SARS emerged in late 2002, an "Asian" pathogen, it provided the perfect opportunity for the global health security regime to test one of its key mechanisms: an enhanced international system of disease surveillance and response. This system, spearheaded by the WHO, monitored the disease's spread within and beyond nations, analyzing data gathered from health information systems and laboratory testing. The global research network took approximately three months to contain the disease.[14] Its swift suppression seemed to represent the unmitigated success of global cooperation.

Amid the lauding of global cooperation and scientific advancement, U.S. public health seized the opportunity to applaud and exaggerate its contributions to SARS containment, while only briefly acknowledging the efforts of its allies. The U.S. CDC took credit for the discovery of the coronavirus, announcing in a March 24 article that "CDC Lab analysis suggests new coronavirus may cause SARS" (CDC 2003a). In reality, the discovery had been the result of the WHO's collaborative Global Outbreak Alert and Response Network (GOARN), which linked together individual disease surveillance and response systems.[15] Moreover, the discovery of the SARS agent (SARS-CoV) had been made in several laboratories in the network simultaneously (Brookes and Khan 2005; Fidler 2005).[16] The CDC article gave the role of the WHO collaborative network only brief mention: "Collaboration among scientists led by the World Health Organization (WHO) is unprecedented." Unnamed were the other major contributing laboratories in Europe, Hong Kong, and mainland China.

The United States repeated its self-congratulatory narrative with respect to the subsequent effort to sequence the virus and learn its genetic characteristics. Canada first sequenced SARS-CoV on April 13, 2003, with the United States following shortly after on the next day. Jerome Hauer, acting assistant secretary for the Office of Public Health Emergency Preparedness, portrayed the United States as leading the sequencing effort in his 2003 testimony during the *SARS: Assessment, Outlook, and Lessons Learned* hearing: "CDC identified the coronavirus within a few short weeks of receiving the first specimens from Asia," followed a few sentences later by "It [SARS-CoV] was successfully sequenced by an international team of laboratories including CDC and Health Canada." The Canadian role was rendered equivalent to the U.S. role. Further, the United States constructed its leadership role by framing other key players as passive recipients of aid: "We have deployed teams of experts and support staff to each of the impacted countries, including Canada, mainland China, Hong Kong, Taiwan, the Philippines, Singapore, Thailand and Vietnam." In fact, Hong Kong, Singapore, and other locations had also completed sequencing and discovered important regional variations (Brookes and Khan 2005). The paternalistic statements of the United States illustrate its attempts to overshadow a sequencing effort that was in fact collaborative and to erase the scientific and medical contributions of other countries.

Thus, the United States ultimately exploited global cooperation to amplify exceptionalist narratives and boost its international clout. Such conceptual practices reflect and uphold U.S. global dominance in an uneven international health system, and bolster U.S. ability to determine health outcomes worldwide. Before elaborating this claim, I outline the history of this uneven system.

The WHO, Neocolonialism, and the Pandemic Turn

Humanitarian medical practitioner-scholar Philippe Calain (2007a, 2007b) and global health politics scholar Sara Davies (2008) have both highlighted the way that the United States and other dominant nations have built global networks that facilitate an unequal system that benefits their parochial interests.[17] Through the WHO, the United States and other wealthy neocolonial powers of the Global North have dominated global health governance, despite the international body's composition of member nations that fall on different ends of the global divide. U.S. public health has played a particularly prominent role in the WHO, where many key U.S. figures hold leadership roles. D. A. Henderson, a U.S. epidemiologist, directed the WHO's smallpox eradication program in the 1960s and 1970s. David Heymann, another U.S. epidemiologist, worked on both the WHO's smallpox eradication program and the U.S. CDC's outbreak containment program in Africa in the 1970s (Lakoff 2015).

Since the end of the Cold War, the CDC has wielded even greater influence over WHO priorities. U.S. public health leaders pushed the WHO to prioritize "emerging infectious diseases," focused on deterring spread from the Global South to the Global North. U.S. health leaders also helped establish the WHO's global network of surveillance centers and reference laboratories (typically private commercial laboratories for pathogen testing and identification) to provide early warning of outbreaks; this data garnered the CDC and other large health institutions information about disease outbreaks abroad before they spread globally and affected, for instance, the United States.

The war on terror afforded the United States and other global powers further opportunity to push the WHO to adopt additional North-centric priorities, namely, WMD event monitoring. This was a notable shift in the international body's purview, and to an issue of much lesser concern for the Global South.[18] The turn to pandemic flu followed soon after, which diverted much-needed resources from the pressing health issues facing most nations in the Global South—that is, diseases linked to poverty and gaps in local health infrastructure (e.g., HIV, tuberculosis, and malaria) (Abraham 2011; Calain 2007a, 2007b).

As mentioned earlier in the chapter, the focus on pandemics revolved around H5N1, which symbolized for the United States the unpredictable threat of influenza strains as well as the specter of "emerging disease" from Asia. The United States responded to H5N1 by creating new global networks to bolster its international pandemic response capacity. In September 2005 at the United Nations General Assembly in New York, Bush announced the International Partnership on Avian and Pandemic Influenza (IPAPI), designed to create "a global network of surveillance and preparedness that will help us to detect and respond quickly to any outbreaks of disease" (Bush 2005). About eighty countries signed on to this international body, which convened a series of meetings beginning January 2006. In the name of "transparency," all who signed on were required to immediately share information and provide flu samples to the WHO should they face an outbreak. The United States further supported global pandemic preparedness by offering both financial and technical assistance for countries to develop capacities for rapid response, lab diagnosis, and surveillance, pledging $434 million to the effort (Bureau of Public Affairs 2007; Crook 2006). In this way, U.S. approaches to disease control shifted toward global efforts that served U.S. interests by generating information—and thus early warning—about diseases before they reached U.S. shores.

The United States also supported the greater authority of the WHO to intervene in the infectious disease response of nations and compel their compliance. This greater authority was encoded in the second of two documents the WHO issued on pandemic preparedness, namely, the revised 2005 International Health Regulations.[19] The culmination of decade-long work to prioritize emerging diseases containment, the regulations built on WHO success in spearheading the global efforts against SARS. Member nations were required to develop the capacity to monitor infectious diseases. In addition, the WHO now possessed greater and broader authority to require member nations to provide "information" about "events that may constitute a public health emergency of international concern" (WHO 2005c). The rubric "events" indicated disease threats and outbreaks that have the potential for serious public health impact and international spread (i.e., pandemics). "Information" included outbreak distribution and trends, morbidity/mortality numbers, but also, potentially, viral samples and genetic sequences—the substrates for making vaccines and other medical treatments.[20] This effectively gave the WHO wider authority to force countries to furnish the international body with data and resources for flu response.

As with SARS, the international surveillance system and its research networks made possible global collaboration, and many nations primarily affected by H5N1 (like China) were contributors and beneficiaries of these efforts. At the same time, the dominance of the United States and other Global North nations positioned them to capitalize on the enhanced control of the WHO over disease resources globally, perpetuating, as I show in the next section, tangible disparities in the international system.

Uneven Surveillance: Discourse and Compliance

China, as one of the main countries H5N1 affected, by the end of 2006 had 14 of the 158 total deaths worldwide, and 22 of the 263 total cases—making it the fourth hardest country hit (WHO 2015). Accordingly, China had great stake in developing adequate flu response and participating in the research networks of the international health system. As poultry outbreaks spread in early 2004, China sent reports to the WHO as part of H5N1 monitoring. China also sent samples collected from infected poultry.

In mid-2005, a senior official in China's Ministry of Agriculture, Jia Youling, stopped sending samples from new H5N1 outbreaks in birds to the WHO and the UN Food and Agriculture Organization. He had discovered that a U.S. team had used samples China shared with the WHO to develop results published in a February 2005 article in the *Journal of Virology* without giving credit to China's Ministry of Agriculture, which had identified the virus. China's grievance was met with a half-hearted apology from the U.S. team—they described it as a mix-up, the lead author calling it an "honest mistake," while also accusing China of hoarding its samples as a rule. The WHO representative to China, Dr. Henk Bekedam, issued a less conditional apology, confirming that the U.S. research team, in failing to acknowledge that China's Ministry of Agriculture had identified the virus they used for their research, was in breach of scientific protocol (Beck 2006; Johnson 2006; Zamiska 2006).

Despite this issue of credit-taking, many U.S. media sources, along with other Western-dominated media sources and the international health community, echoed the U.S. research team's accusation that China in general hoarded information. In reference to China's dispute with the WHO, a July 2005 *Washington Post* article described China as refusing to hand over any information other than the most basic outbreak data (other information includes, for example, specifics about the extent of infection or genetic analyses of the strain). The article characterized China as withholding: "The [Chinese] government has yet to respond to a . . . request by international health experts," and "Chinese officials did not respond to written requests for comment by the Washington Post." The article gave no context for China's actions other than cageyness: "U.N. officials and independent scientists said they were reluctant to publicly discuss their frustrations with China for fear the government would shut them out of the country. But officials and researchers said they were dismayed with the government's secrecy, especially after China ran afoul of international agencies for its response to the SARS epidemic that began in 2002" (Sipress 2005).

In invoking a secretive, obfuscating Chinese central government, the article implied that the Chinese government could not be trusted in health endeavors. The characterization of China's H5N1 response in the static terms of a caricatured SARS response— namely, a cover-up[21]—is problematic in that China responded vigorously to the H5N1 outbreak, swiftly sealing off, vaccinating, and later culling infected birds to mitigate the pathogen's ability to infect humans, as well as ramping up medical supplies, training, and funds once it had. In fact, China had boosted spending on health after SARS, including on infectious disease and especially on HIV/AIDS. China has continued to address important health issues, implementing in 2008 the first of several policies for universal medical insurance, as well as pensions, medical leave, and other health systems reforms. Even so, Western scholars often disparage China's increased attention to disease response as a desire to boost its international image—a "saving face" trope, rather than a desire to improve the country's health (Wishnick 2010).

This image of China is rooted in longstanding Western portrayals of China as backward and diseased,[22] as well as more recent portrayals of China's government as secretive, stemming from the Cold War, when China aligned with the Soviet Union (Kim 2010). The resurgence of denigrating tropes of China at the turn of the twenty-first century occurred in the context of China's rising military and economic influence in the world, and U.S. desire to limit that influence. Although China's power was nowhere near on par with that of the United States—especially its military power—the United States has maintained what historian Vijay Prashad calls "Western fantasies of Chinese domination" (Prashad 2017, 2541).[23]

It was in this geopolitical milieu that SARS emerged, and the Chinese government's handling of SARS met with enormous U.S. scrutiny. U.S. public health and news media portrayed Chinese public health practice as in shambles and Chinese culture as possessing, in the words of one journalist, a "casual attitude toward health" where "men and women enthusiastically spit in public" (Lynch 2003). They attributed the emergence of SARS in China to exotic and savage food preparation and consumption practices (D'Arcangelis 2008)—a place where humans live "cheek-by-jowl" with a variety of animals that "walk in and out of their houses" (Pearson et al. 2003).

Few stories remarked on China's relatively low reported mortality rate from SARS—at approximately 6.5 percent—much lower than that reported worldwide (9.6 percent) and significantly lower than, for example, U.S. neighbor Canada (17 percent) (WHO 2003c).[24] Nor did U.S. coverage of China mention the extreme range of scholarly viewpoints on how China dealt with SARS and on the state of Chinese public health at the time: while some criticized China for withholding information about the outbreak for too long and putting economic considerations ahead of public health, others praised Chinese public health response as relatively quick in the face of a new disease.[25] In effect, SARS afforded the United States the opportunity to reinvigorate the specter of a diseased Orient[26]—a story that only intensified with the advent of H5N1.[27]

CDC director Julie Gerberding described Asia as "the perfect incubator" for pandemic flu (Manning 2005). Similarly, Senate staff director David Dorman opened the *China's Response to Avian Flu* hearing by referring to China as "one of the prime incubators for a potential human influenza pandemic." The reduction of China to solely a disease generator went hand in hand with U.S. reliance on national barriers as a means of disease control, despite their inefficacy.[28] Representative Tom Lantos reflected this approach in his 2005 testimony during the *National Pandemic Influenza Preparedness and Response Plan* hearing: "This pandemic, if it comes, is most likely to come from Asia, it is most likely to come via San Francisco, Los Angeles or other ports of entry." Such a focus failed to promote a much more efficacious approach, namely, cooperation with Asia's disease control efforts—overlooking the region's demonstrated capability at collaborative health management.[29]

A *New York Times* article titled "Scientists Hope Vigilance Stymies Avian Flu Mutations," which described U.S. scientific and health efforts in flu response, contrasted the United States and China in a pair of photos (McNeil 2007).

FIG. 7. "Bird flu: Dr. Mitch Cohen and Julie L. Gerberding of the Centers for Disease Control and Prevention discussing pandemic flu. [Bottom], a poultry market in China's Guangdong Province." From Donald McNeil, "Scientists Hope Vigilance Stymies Avian Flu Mutations," *New York Times*, March 27, 2007, F1. (Top: Erik S. Lesser/The New York Times/Redux. Bottom: AP Photo/Color China Photo.)

The top photo in figure 7 depicts two U.S. professionals—CDC Director Gerberding along with another doctor; the bottom photo in figure 7 depicts a Chinese man surrounded by a flock of poultry in a dark, dank-looking room. The choice of juxtaposing a Chinese farmer with U.S. doctors reproduced a view of China as a generator of disease. In contrast, the United States was portrayed as the epitome of science and health—we see only experts, far removed from the source of disease. These depictions, of China as disease producer and the United States as model of flu readiness, were the denouement of the Orientalist imaginary that had gained traction since SARS.

Chandra Mohanty (1984) describes how knowledge production can function as "discursive colonization," a "mode of appropriation and codification of 'scholarship' and 'knowledge' about [a non-Western subject] by particular analytic categories employed in specific writings on the subject," serving to enact "a relation of structural domination, and a suppression—often violent—of the heterogeneity of the subject(s) in question" (333). Because U.S. discourse about China during the H5N1 response flattened China into an impediment to global health, it ignored a much more complex picture. China was submerged into the category of "non-West" and by association reduced to the downtrodden and technologically backward. The recognition that many non-Western nations have been colonized and exploited by the West (which has contributed in large part to their ailing public health infrastructures) should not preclude a more nuanced discussion of each nation's actual public health capacity.

These flattened portrayals, moreover, diminished China's governance capacity. To return to the controversy over China's sample sharing stoppage in response to U.S. credit stealing, the ensuing negative press effectively acted as compliance enforcement. Press coverage produces strong incentive for countries to comply with the international system; as Briggs and Nichter (2009) state: "If countries are likely to be 'outed' by such [international surveillance] networks, it behooves them to report incidents in a timely manner and to be seen as good global citizens instead of selfish agents trying to protect their self-interests in trade, tourism, institutional politics, political alliances, and national images"; consequently, "[some] countries are rewarded in the press for being 'good global health citizens,'" while others are "blamed as bastions of unhygienic subjects" (197). Discursive mechanisms can, in fact, force compliance as strongly, if not more so, than formal compliance mechanisms (especially since the revised 2005 International Health Regulations guidelines lacked any formal compliance mechanism to require member nations to provide information and samples).[30]

Moreover, in a neocolonial international system, global pressure is not evenly applied. While the international press and health communities met China with charges of obfuscation and narratives of unhygiene, they largely left the United States alone. In fact, the WHO repeatedly accused China of delaying reports of H5N1 outbreaks, even though China reported regularly to the WHO during and after avian flu outbreaks (Davies 2012, 603).[31] Thus, the international system and its Global North leaders can browbeat less powerful, Global South nations to comply with the North-centered international system, regardless of how this might diminish their ability to implement disease control in ways that best protect their populations. Indeed, this pressure seems to have worked in China's case—China's Ministry of Agriculture resumed sending samples a year later.

Extractive Biocolonialism: Germs as Resources

Flu samples are the crux of pandemic preparedness—a key resource for health, research, and profit. Since the 1950s, nations had been providing the WHO's Global Influenza Surveillance Network (GISN)[32] with flu samples from patients: national laboratories in the network collect the samples, then share information and strains for further analysis with the network's international centers, as well as with researchers and pharmaceutical manufacturers trying to develop vaccines and other medical products to combat and/or treat influenza such as antiviral drugs and diagnostics (Bresalier 2012; Kitler, Gavinio, and Lavanchy 2002; Vezzani 2010). By the turn of the twenty-first century, this "virus-sharing system" included over one hundred national laboratories and four international centers (located in the United States, the UK, Japan, and Australia).

The U.S. CDC and the European CDC play prominent roles in collecting samples and analyzing data for the GISN, and they also contract with major vaccine manufacturers so they have first access to the vaccines produced. The vaccine production industry—a global infrastructure of vaccine factories, laboratories, and pharmaceutical companies—is concentrated in the United States, Canada, Australia, and Europe (Fedson 2003). For this reason, wealthy countries possess the research and development capacity to produce vaccines and other treatments, and control who can access them.

As global security scholar Shawn Smallman (2013) notes, the international health system of virus sharing embodies a neocolonial relationship: it perpetuates the dependency of poorer countries impacted by disease on wealthy ones, which secure the needs of their own peoples while "keeping the governments of poorer states as supplicants" (22).[33] In practice, the virus-sharing system makes affected nations give up their strains so that they can serve as the substrate for developing treatments primarily accessible to wealthy nations. This system unconscionably leaves nations without the means to purchase them at the mercy of drug donations gifted by wealthy nations and pharmaceutical manufacturers.

While pandemic flu preparedness did not birth this skewed system, it exacerbated its inequities. In 2003, the GISN turned its attention to variants that might signal the start of a pandemic, and this move stimulated the demand for samples of potentially pandemic strains as well as the production of vaccines and antivirals against them. As demand intensified, the flu strains of afflicted nations became an increasingly precious commodity sought by nations in the Global North and their pharmaceutical partners. In the United States, where H5N1 did not appear (in birds) until much later in 2014, federal flu plans had nevertheless devoted significant attention to vaccine production as the magic bullet in flu response, and enlisted the vast pharmaceutical infrastructure by offering increased funding and other incentives.[34]

This infrastructure in turn depended on the acquisition of samples from afflicted nations—China, as already mentioned, and also Indonesia. The H5N1 strains that hit Indonesia were some of the deadliest, with the nation experiencing 58 of the 158 total deaths worldwide, and 75 of the 263 total cases, by the end of 2006 (WHO 2015). Indonesia became a major focus of the international health system and a key site where its inequities materialized. Moreover, like China, Indonesia resisted the inequitable international system.

Indonesia had been providing samples to the GISN since finding its first human case in July 2005. But in December 2006, Indonesia stopped. Indonesian health minister Siti Fadilah Supari explained these actions, in her public statements and publications,[35] as a refusal to engage in a skewed international system of "virus sharing." This included issues of credit-taking similar to what had befallen China: in April 2006, results of laboratory analyses involving viruses from Indonesia were presented at various international meetings without permission or with last-minute notification, and papers were written that added researchers from Indonesia as co-authors only at the later stages of writing (Sedyaningsih et al. 2008). Supari also highlighted how the international system negatively affected Indonesia's health, when the Indonesian government could not obtain antiviral oseltamivir (Tamiflu) in late 2005 to treat initial cases of H5N1 because wealthy countries—which had yet to have any cases—had bought up the supply for their preparedness stockpiles (Supari 2008).

Indonesia's inability to access necessary treatment came to a head at the end of 2006, when the Indonesian Ministry of Health discovered an Australian company's plans to develop a vaccine against H5N1—using a strain Indonesia had provided the WHO system, and without Indonesia's permission (Fidler 2008; Sedyaningsih et al. 2008). This violated WHO guidelines requiring the donating countries' permission to develop vaccines.[36] The

Australian drug company, CSL, later admitted to using Indonesian avian flu strains to develop a trial vaccine, while also insisting that it had no obligation to compensate Indonesia or guarantee the nation access to the vaccine (Lakoff 2010). In an article published in 2008, the Indonesian Ministry indicted the virus-sharing system, stating: "Disease affected countries, which are usually developing countries, provide information and share biological specimens/virus with the WHO system; then pharmaceutical industries of developed countries obtain free access to this information and specimens, produce and patent the products (diagnostics, vaccines, therapeutics or other technologies), and sell them back to the developing countries at unaffordable prices" (Sedyaningsih et al. 2008, 486).

The ministry showed that the system's inequity lies beyond the fact that Indonesia lacked the manufacturing capacity and purchasing power of the rich countries to acquire these high-demand treatments. The problem also lies in the way the system demanded viral resources without compensation for afflicted nations. The wealthy countries of the Global North could access viral resources from the Global South for free, produce treatments derived from these resources, and then hoard them.

Scholars and activists have used the term "biocolonialism"[37] or "extractive biocolonialism"[38] (Whitt 2009) to describe how the Global North appropriates biological riches (plants, animals, bodies, tissues, genes, etc.) from indigenous peoples, the poor, the marginal, the weak, the subjugated, and the genetically distinct. Well-known examples include U.S. researchers and corporations patenting medicinal plants from abroad, such as neem and ayahuasca, and the Human Genome Diversity Project's commodifying the cell lines of indigenous peoples. This biocolonial system is, moreover, the enduring legacy of colonial extraction of land, labor, knowledge, and bodies, which has been updated in the contemporary context of a knowledge-based economy, privatization, and the patent system (Hawthorne 2007; Thacker 2006). An interconnected cast of powerful characters—Global North governments and transnational corporations—are the beneficiaries of biocolonialism;

they acquire an array of benefits, from health and scientific advancement to reputation enhancement and profit.

Diseases may not seem like they fall under the rubric of "biological riches," since they are destructive entities that humans generally seek to stamp out.[39] Yet, like plants and genetics, disease—the viral samples and other materials from which disease data can be gleaned—do constitute a resource, one procured from the Global South for the advancement of the Global North. Thus, the barriers Indonesia faced to obtaining the medications it urgently needed, despite the fact that these medications originated from its viral resources, were a product of the biocoloniality of the virus-sharing system.

In addition to creating tangible vulnerabilities for Global South locales such as China and Indonesia, biocolonial extraction of samples and data imposed the systems and frames of the Global North. North-centric regimes of property and ownership forced afflicted nations to "share" viral data even while they enabled corporations to monopolize vaccines and their distribution. The ability to patent, and thus hold exclusive rights over viruses and the products derived from them, turned pathogenic flu samples into a form of biocapital that pharmaceutical industries based in the Global North sought to acquire to develop lucrative products for sale. A report on patenting trends shows that in 2005 patent applications skyrocketed for H5N1 genetic sequences, vaccines, treatments, and diagnostics—from single digits in prior years to over forty by 2007. The countries in the Global North comprised the vast majority of patent holders. Over half (53 percent) of the patents filed in 2007 originated in the United States (pharmaceutical companies, state-funded labs, government agencies such as the CDC); 27 percent were from Europe; and 4 percent were from Australia (Hammond 2009). Thus, patent holders largely based in the Global North determined who to sell the treatments to. Unbound by the health demands of afflicted nations, government agencies in the Global North could focus on preventative measures for diseases they were not even afflicted by, and private corporations could make decisions based on profitability.

Indonesia's challenge to the biocolonial system did not end with its refusal to donate flu samples to the WHO. Indonesia would propose a profound restructuring of the flu sample sharing system with the goal of affording originating nations (nations where the flu samples came from) greater influence over the process. The Indonesian Ministry of Health challenged the way the WHO interpreted the 2005 International Health Regulations (known as IHR 2005), the legally binding guidelines through which the WHO enacted infectious disease management. They pointed out that the WHO was acting on the assumption that the IHR 2005 required that biological samples be shared, even though the IHR 2005 did not specifically require nations to share biological samples, only information: "Public health information and biological substances are 2 independent concepts" (Sedyaningsih et al. 2008, 484–485). Whether information could be interpreted as encompassing samples became subject to vigorous debate.

Indonesia moved the needle further, pushing for originating countries to attain increased control over their samples, drawing on the Convention on Biological Diversity, an international treaty that recognizes countries' sovereign control over their biological resources (plants, organisms, genetics) and grants them the authority to determine access to these resources.[40] The ministry argued that viral strains found in their countries were therefore protected: "Hence, countries have the right and authority to decide whether to share their specimens with the WHO system or not, depending on their own judgment" (Sedyaningsih et al. 2008, 485). This challenged a key foundation of the biocolonial extraction apparatus, namely, WHO ability to supersede national autonomy in the name of global health security.

Indonesia sedimented this shift by proposing several changes to the system. First, Indonesia wanted to share virus samples only with parties that signed material transfer agreements (MTAs), which govern the conditions of transfer of biological materials and usually stipulate how the recipient may use the material, essentially maintaining control with the original provider. Ironically, the WHO required just such a condition for its own samples: once samples were acquired from originating countries, and at no charge, the WHO required parties it shares samples with to sign MTAs with the WHO. In this way, Indonesia utilized existing proprietary regimes to try to redistribute power to originating countries.

Indonesian officials also pushed for equity in terms of health outcomes: they wanted vaccines available to all countries at risk of being affected at a minimal price they could afford (Sedyaningsih et al. 2008). They centered the principle of "benefit sharing"—that countries where the viral strains derived from would receive some benefits, whether that entailed wealthy nations, pharmaceuticals, or the WHO providing and sharing resources (labs, etc.) to originating countries or guaranteed access to the vaccines and other products derived from the viral strains. In her 2008 book, Supari elaborated: "Benefits sharing is a consequence of virus sharing, which instead of a charity from the developed country to the country where the virus originated, it is the right of the latter" (116–117 [text reproduced in Elbe 2010]). In replacing the concept of virus sharing with benefits sharing, the Indonesian Health Ministry decentered hegemonic regimes of property and the patent system, pivoting instead to the rights due disease-afflicted nations.

In the midst of these statements and actions, Indonesia garnered the support of other nations in the Global South. At the WHO's Executive Board meeting in January 2007, the Thailand representative also underlined the inequities of the virus-sharing system and the catastrophic consequences for the health of poor nations: that "[we] are sending our virus [samples] to the rich countries to produce antivirals and vaccines. And when the pandemic occurs, they survive and we die. . . . We are not opposed to the sharing of information and virus [samples], but on the condition that every country will have equal opportunity to get access to vaccine and antivirals if such a pandemic occurs" (Fidler 2007). Indonesia had, in effect, made room for the Global South's counternarrative that need, rather than wealth, should determine access.

The international scope of the dispute was on full display at the May 2007 World Health Assembly, which is the policy decision-making body of the WHO. Twenty countries in the Global South put forth a resolution supporting Indonesia's position on national sovereignty over viruses and the principle of benefit sharing. They called for a new international framework for sharing avian influenza viruses; to review the existing WHO research system; and to prioritize the manufacture and availability of vaccines in developing countries. They argued that any vaccines, diagnostics, antivirals, and other medical supplies arising from the use of the virus and parts thereof must be available at an affordable price and in a timely manner to developing countries, particularly those under the most serious threat or already experiencing the pandemic threat (WHO 2008).

Ultimately, Indonesia and its allies successfully instituted change. In the many subsequent intergovernmental meetings and working groups to address the dispute (the first of which convened in November 2007), the Global South pressed its concerns about the virus-sharing system, eventually forcing both the United States and the WHO to make substantive concessions that granted more control to originating countries and more compensation from recipient countries.[41] In the end, the World Health Assembly passed the Pandemic Influenza Preparedness Framework (adopted on May 24, 2011), which created a vaccine stockpile for developing countries and new rules regarding influenza virus sample sharing. The resolution established "partnership contributions," whereby pharmaceutical manufacturers that use the GISN (which was at this time renamed the GISRS, Global Influenza Surveillance and Response System) had to contribute annually.

The agreement also recognized national sovereignty claims over viruses; established a tracking system whereby contributing nations could track the samples they donated to the WHO; and established two forms of MTAs, one of which required vaccine manufacturers acquiring materials through GISRS to donate a portion of their vaccine production to the WHO or developing countries (WHO 2011). The Global South did not achieve all its aims—there were no limitations on patenting, and left out was a proposed endowment that would have ensured that the stockpile was large enough for the needs of developing countries (Aldis and Soendoro 2014; Smallman 2013). However, the agreement represented a significant step toward equity in the international health system.

Collective action had brought about a profound restructuring in global health governance, allowing significantly less powerful actors to be relatively successful against more powerful ones. Discourse-making, moreover, was vital to their accomplishments. The strategic displacement of "virus sharing" to "benefits sharing" destabilized the entire discursive scaffolding of "global health security." Benefits sharing reversed the construction of vulnerability: if the existing view was a Eurocentric one premised on the notion that the Global North needed to be protected from the diseases of the Global South, the new paradigm highlighted that it was the Global South that was made vulnerable through the North's biocolonial extraction of disease data and samples.

This was a hard-won paradigm shift, as illustrated by the U.S. pushback in the struggle to obtain MTAs for originating countries. The United States had a strong stake in a system that gave global powers prime access to the data and samples furnished to the WHO. Accordingly, the United States rejected the Global South's attempt to redistribute power through the MTA regime. At the World Health Assembly meeting in May 2007, the United States presented its draft resolution, which argued that adding MTAs would hinder the unfettered, immediate transfer of materials, and thus negatively impact vaccine production (Franklin 2009; Khor 2007; WHO 2008). In essence, the United States stuck to the notion that vaccine production was the sole goal, ignoring the crucial step that followed—how those vaccines would be distributed.

The paradigm shift to benefits sharing also challenged another fundamental premise of global health security—that international authorities should supersede national ones. Benefits sharing repositioned national sovereignty over biological resources as the key mechanism for giving impacted nations control over their health outcomes. The focus on national sovereignty was, in the terms of sociocultural anthropologist Aihwa Ong (2008), a strategic insertion of "the nation as a scale of ethical exception to the global commodification of health" (126). The concept of national sovereignty prioritized the well-being of peoples actually affected by H5N1, and interrupted the predatory extraction of samples from the Global South. Thus, national sovereignty was articulated not as selfish or provincial, but as key to ensuring the health of those most impacted.

The discourse of national sovereignty also chipped away at the rhetorics of security that had helped justify pandemic preparedness as a *security* imperative. It was certainly a difficult battle: the discourse of global health security had not only diminished national autonomy, but also coded nations such as China and Indonesia, who asserted their authority, as security threats. At a March 2007 WHO meeting held in Jakarta, for example, a WHO press release called Indonesia's withholding of viruses a "threat to global public health security" (WHO 2007). This married false universalism (a disaggregated globe) with alarmist security rhetoric to sideline the equity issues that had been highlighted by Indonesia and its allies.

U.S. government officials, along with much of the U.S. popular press, had also frequently deployed these security frames in response to Indonesia's actions. Prominent public health journalists Richard Holbrooke and Laurie Garrett, in their denouncement of Indonesia's position, published an article in August 2008 in the *Washington Post* that described Indonesia's failure to provide viruses as posing a "pandemic threat to all the peoples of the world." The authors followed this with further security discourse: "Disturbingly, however, the notion [of viral sovereignty] has morphed into a global movement, fueled by *self-destructive, anti-western sentiments*" (Holbrooke and Garrett 2008; emphasis added). This characterization invoked the trope of the West-hating terrorist, but also the post-9/11 frame that was its complement—U.S. vulnerability, here aimed at cowing Global South nations that refused to comply with the North-led international system. The authors ended their article by calling on Indonesia to conform to the demands of "globally shared health risk," further coding Indonesia's challenge to the international health system as a global threat and upholding the myth that the international pandemic preparedness regime—and its fortification—benefited all nations.

Indonesia finished what China had started when the latter briefly staged a flu sample sharing stoppage. The shift to benefits sharing and national sovereignty, and away from the rhetorics of global health security, diminished both the discursive and the institutional power of the North-led international system. Indonesia, along with its allies, interrupted the biocolonial extraction of resources from the Global South and altered the system itself, institutionalizing a redistribution of power to give the Global South a greater role in global health governance.

Conclusion: Transnational Approaches to Justice

Infectious disease control has been marred by a neocolonial network of international health organizations (most prominently the WHO), wealthy nations in the Global North, and pharmaceutical companies based in these nations, all of which engage in extractive biocolonialism with the Global South, acquiring their data, specimens, and other disease resources. The turn to securitization initiated by the United States exacerbated the problems with this existing system. Indonesia and its allies, like the nurses contesting the National Smallpox Vaccination Program described in the last chapter, successfully pivoted dominant health discourse, questioning *whose* health matters. In both cases, actors outside the centers of power made room for equity and anticolonial discourses that challenged the dominant security frame, pushed beyond disingenuous universals, and engendered careful examination of how risk is offshored onto particular regions and nations.

I hope, in delineating this inequity in the international health system, the hegemonic narratives of U.S. stewardship, and the counter-hegemonic foothold that Indonesia achieved, to cultivate awareness of more avenues to truly cooperative and egalitarian global health initiatives that serve those whose health is most vulnerable. My analysis has aimed to contribute to the displacement of the misleading "we're all in this together" mantra of global health security, as well as the paternalism of U.S. exceptionalist rhetoric that positions the United States as best equipped to lead global health efforts.

Further, my work seeks to engage with other critics of the post-9/11 preparedness regimes, particularly those who have focused on the diversion of much-needed resources from known, common, and everyday illnesses and afflictions like HIV and TB, which predominantly affect low-income and marginalized communities. I would like to push for a transnational analytic that connects the health access of marginalized groups within the United States to marginalized groups worldwide who are fighting for their health priorities to matter. Such a frame, which focuses on the connections between the U.S. preparedness complex's effects within the United States and its effects globally, might help build even more solidarities in dismantling U.S. bio-imperialism and the neocolonial global health system in which it is embedded.

Epilogue

Repurposing Science and Public Health

Our movement lives and dies with the broader left; technical
knowledge alone never delivers justice.
—Science for the People, "The Dual Nature of Science,"
April 12, 2018

Nurses take a sacred oath to care for anyone who needs help, and
inherently reject intolerance, racism, hate, and bigotry.
—National Nurses United, "RNs of National Nurses United Say
Racism, Xenophobia Combined with Lax Gun Control Laws at
Root of Mass Shooting Epidemic," August 4, 2019

In the first decade of the twenty-first century, U.S. bio-imperialism
found purchase in the operations of the war on terror in health and
bioscience. The U.S. national security apparatus mobilized racial
and gendered tropes of the bioterrorist Other and national vulnerability, as well as narratives of technoscientific progress and caregiving. These helped the U.S. state revamp its relatively dormant
war-making capacity in germ weapons and enhance its position in
global health networks to obtain germ resources and reinforce the
public face of U.S. global stewardship. U.S. (and global) elites—
dominant social groups as well as corporate entities—benefited,
while Arabs, Muslims, and East and Southeast Asians bore the
brunt of bio-imperial targeting, and scientists, lab workers, and
caregivers shouldered its collateral costs.

These imperial machinations in the germ realm did not expire with the Bush administration in 2008, nor with the official end of the war on terror on May 23, 2013. In this final chapter I tackle the enduring features of the post-9/11 preparedness regimes—through the Obama presidency and the first two years of the Trump presidency.

Against Militarism and Elitism in Science: Public Accountability, Mass Movement

Under Obama, the United States continued its policy of unilateralism with respect to the Biological Weapons Convention (BWC), with Secretary of State Hillary Clinton announcing at the seventh review conference in Geneva on December 2011 that the United States would still not endorse verification protocols. Secretary Clinton cited difficulties in assessing biological research as offensive versus defensive (this differed from the Bush administration's rationale only in that the latter had also cited industry's proprietary interests) (White House 2011). The Obama administration insisted that the treaty was important and, in lieu of a formal verification regime, put forward a twenty-three-page *National Strategy for Countering Biological Threats*. The strategy outlined indirect measures to prevent offensive bioweaponry, such as laboratory security and other Bush-era biosecurity measures focused on terrorism (National Security Council 2009).[1] The Obama administration also maintained the enlarged U.S. biodefense research enterprise of its predecessor, though the Obama administration scaled back the riskier pursuits of Bush-era preparedness such as threat characterization research (the production of potential weaponized pathogens to learn more about them).

Under Trump, biodefense funding decreased in comparison with the preceding years under Obama, although to levels still far above what they were pre-9/11 (Watson et al. 2018). This downward shift was the result of more, not less, militarism; under Trump, Homeland Security funding allocations were primarily directed to aggressively amplifying U.S.-Mexico border patrol and

Immigration and Customs Enforcement (ICE) raids across the country (Machi 2017). Trump, moreover, opted for boosting U.S. nuclear power, signing the Nuclear Energy Innovation Capabilities Act on September 28, 2018, which eliminated financial and technological barriers to the development of advanced reactors in the United States. Biodefense defunding notwithstanding, the United States has not given up its power in the biological (and chemical) weapons arena: on April 13, 2018, it launched air strikes against Syrian facilities linked to the production of chemical and biological weapons.[2]

In order to sever science and public health from U.S. empire—with its trappings of militarism, racism, and sexism—those centrally involved and directly impacted must play a key role. This book has recounted that U.S. scientists failed to resist the Bush era specter of bioterrorism. As the Bush administration expanded the biodefense industry and secured the research therein from "restricted persons"—that is, foreign nationals hailing from "state sponsors of terrorism" and other groups criminalized under the USA PATRIOT Act—bioscience became yet another arena stoking a racist, nativist post-9/11 imaginary. Scientists who objected to the new restrictions on accessing and transferring research materials did so largely because such restrictions hampered their "scientific freedom." Their actions reflected the ideology that science is neutral and value-free, and that scientists should be free to pursue any and all research for the sake of technoscientific progress, while taking no responsibility for the consequences and political dimensions of their work.

Criticism of science's role in the war on terror came instead from outside the scientific establishment. Progressive journalists and organizations concerned with science, such as biodefense watchdog group the Sunshine Project and bio-artist collective Critical Art Ensemble (to which Steve Kurtz belongs), were vital to holding the biodefense industry accountable. They critiqued the role of science in U.S. war and empire, but also the fact that the U.S. public could obtain little information on these projects (the increasing purview of the post-9/11 U.S. national security apparatus was directly proportional to decreasing public transparency). These groups made the clandestine activities of the U.S. biological warfare industry and its hazards publicly visible. Pushing past the barriers of technical jargon and the secrecy shrouding security practices, they trespassed into realms typically reserved for the technocratic elite. In doing so, they accomplished an important first step toward public accountability.

In 2014, the activist organization Science for the People revived its sixties-era mission to "mobilize people working in scientific fields to become active in agitating for science, technology, and medicine that would serve social needs rather than military and corporate interests."[3] The organization had emerged out of the antiwar movement in 1969, after science students at MIT organized a moratorium on war research, leading to protests at other universities. Out of that activism came Scientists and Engineers for Social and Political Action, which later changed its name to Science for the People. The organization critiqued the mainstream scientific establishment's complicity in war, as well as sexism, racism and capitalism. Members waged protests and boycotts, and until 1989 published more than a hundred issues of a bimonthly magazine, focusing on topics such as biological determinism and farm worker mechanization.

In its current twenty-first century form, Science for the People continues to organize scientists, along with activists, students, and scholars interested in science, to take control of the scientific agenda and direct it toward "a more humane and liberating vision of science in society" (Science for the People n.d.b). The organization's working groups agitate against the status quo of military, corporate, and elite interests directing science.[4] They connect to radical left movements, from Black Lives Matter's opposition to state violence against black bodies to immigration justice organizations' fight against ICE's dehumanizing, violent actions at the border ordered by Trump's openly racist, xenophobic regime. Science for the People has engaged in solidarity pickets, most notably with tech workers calling for companies Microsoft, Google, and Amazon to stop supplying technology platforms to ICE

(Science for the People n.d.c). It has also held events in support of the Green New Deal, legislation that foregrounds the differential impacts of climate change on women, indigenous peoples, peoples of color, and the Global South (Science for the People n.d.d).

This renewal of sixties-era mass, organized movement holds the promise to repurpose science and technology for social justice. In rejecting the ideology of unregulated technoscientific progress and attempting to decouple technoscientific practices from their imbrication with militarism, racism, and sexism, theirs is a radical vision that holds scientists and tech workers accountable for the consequences of their labor—research and development practices should be rigorously evidence-based as well as socially meaningful. It entreats scientists and tech workers to reconsider the ends to which they ought to put their labor power: "Knowledge is won with our labor and can be used to advance common goals" (Science for the People 2018).

From Local to Global Wellness: Building Intersections, Practicing Solidarity

When Obama inherited the biodefense budget of his predecessor, he conserved its levels, but rerouted funding from biodefense to prevention of naturally arising diseases (Tucker 2010).[5] Unlike bioweapons attacks, infectious disease remained a very real occurrence. The H1N1 "swine" flu emerged in April 2009 in Mexico and the United States (California).[6] Less than two months later, the World Health Organization (WHO) labeled it a pandemic. The Obama administration met it with the pandemic preparedness regime set in motion under Bush, but scaled back the contested role of border controls—foregoing travel restrictions—and de-emphasized militarization as a means to secure public compliance (*The 2009 Influenza Pandemic* 2009). The Centers for Disease Control and Prevention (CDC) turned to equitable vaccine distribution, prioritizing populations most at risk for severe illness (pregnant women, caregivers for young children, healthcare workers and

ER personnel, children and young adults, and people with health conditions) (CDC n.d.c).

U.S. efforts experienced some stumbles—H1N1 vaccine stocks ran out as pharmaceutical production slowed. Even so, flu vaccines are not a perfect solution—strains mutate and make vaccines less effective or obsolete. U.S. dependence on vaccines as the primary response to flu, moreover, can distract from addressing underlying structures that cause potentially pandemic human flu strains to emerge. Industrial factory farming has produced large-scale animal outbreaks—the densely packed, highly susceptible animals then pose a threat of infection to their human handlers (Greger 2006); these farms also create large quantities of animal waste that frequently contaminate nearby waters with viruses and other germs that humans may consume, which can lead to strains that can then infect and transmit among human populations (Singer 2009).

In addition to vaccination being an inadequate prevention strategy, as long as health interventions rely heavily on vaccines and pharmaceutical production, they remain entrenched in the agendas of global corporate institutions. Yet, we must look to the incredible strides Indonesia and its allies in the Global South made when they banded together during the H5N1 epidemic to drastically transform a flu virus-sharing system that was both deeply beholden to pharmaceutical corporations and structured around the health priorities of the Global North. They bucked the profit-driven, neocolonial virus-sharing system through a combination of systemic analysis, strategic resource-leveraging, and alliance-building. Through these sustained activities they built a global flu system that foregrounded benefit sharing and the health of nations most impacted by flu, which culminated in the 2011 Pandemic Influenza Preparedness Framework adopted by the WHO. Although broader inequities in pathogen sharing have remained, as North-based pharmaceuticals continue to find ways to acquire the South's disease samples without providing adequate compensation,[7] the reformed global flu regulations nevertheless constitute an extraordinary success for health equity.

The way that national governments in the Global South were able to redistribute the international balance of power in flu preparedness serves as a valuable model for achieving global justice. But social movements and movement-based organizations remain pivotal in effecting transnational change. The Third World Network, an independent nonprofit international research and advocacy organization based in Malaysia, supported Global South governments' efforts to rebalance the WHO global flu system;[8] they were also vocal advocates for indigenous groups, which face marginalization in the Global South and North alike in access and benefit-sharing systems because these systems prioritize state sovereignty—rather than indigenous people's sovereignty—over biological resources (Saez 2018).[9]

The nurses' organizations active in countering the 2003 National Smallpox Vaccination Program offer a U.S.-based exemplar of localized change-making that connects to global justice. When they challenged the vaccination program, calling it out as more harmful than healthful, they built on years of organizing through professional associations and unions for safer working conditions and better pay within male- and corporate-dominated medical establishments in the United States (Reverby 1987). And when they charged the Bush administration with devising the vaccination program to boost its war-mongering agenda against Iraq, they connected with broader struggles against U.S. empire, acting in solidarity with the Iraqi people.

In 2009, the California Nurses Association, the Massachusetts Nurses Association, and the United American Nurses joined to form the broader National Nurses United (NNU), becoming the largest organization of registered nurses in the United States.[10] They have continued to fight in solidarity with progressive movements globally. Under Obama they were at the forefront of fighting for universal health care (Healthcare-NOW 2009), and under Trump's austere health care vision they have continued to lobby for Medicare for All (National Nurses United 2012).[11] They have also been quick to take action against the Trump administration's perpetration of humanitarian and health crises: they have worked at the U.S.-Mexico border in refugee support, advocated against the gun epidemic, denounced the administration's tacit support of white male/white nationalist violence, and promoted the global movement for climate justice (National Nurses United 2014, 2019a). The world's health problems remain daunting, as many governments around the world continue to disinvest in disease prevention and health infrastructure, as well as over-rely on vaccines and other pharmaceutical treatments—made worse by corporate monopolies and widening access disparities. I continue to heed the organizations and movements working to disentangle health care from profit and militarism, and how they ground their struggle in coalition building with other movements against injustice. These groups, I believe, will persist in making significant inroads on a structural level—and on a global scale—to transform health care into a comprehensive, equitable, and accessible public service.

Under the blatantly misogynist, racist, nativist, and classist regime of the Trump administration, increasing numbers of people in the United States have joined the struggle for social justice, with some even questioning the longer history of U.S. state violence via policing, border control, criminalization, and the destruction of families. Understanding the deep roots of U.S. injustice is key to crafting better futures that prioritize equity and the lives of the majority—not just in the United States but globally. A retooling of science and public health must be part of this broader social justice agenda. We would do well to follow the lead of progressive health workers and scientists, alongside biodefense watchdog groups and transnational health equity organizations, all of which have laid the groundwork for this way forward.

Acknowledgments

The seed of this book came from research begun over a decade ago. After generating several articles at the intersection of gender, race, science, health, and national security, in 2015 a new theme crystallized for me: bio-imperialism. It was how I thought of the legacy of the war on terror in the biosciences and public health, and I believed it remained highly relevant. My first task was visioning the book's overall purpose and structure. My idea exchange with Emily Cheng, who was also at work on the early stages of her book, was invaluable: we brainstormed together, traded half-baked thoughts, and generally shared the journey of the first book. Once I dove into the writing phase, my writing group was essential in helping me iron out my ideas and polish my prose—chapter by chapter; a big thank-you to Lisette V. Balabarca-Fataccioli, Silvia Mejía, Sudarat Musikawong, Oscar Pérez-Hernández, and Barbara Sutton.

I have also worked with two wonderful editors: Claudia Castañeda provided me excellent feedback on early chapters; Colleen Jankovic gave me amazing feedback at the manuscript's end stages—her insights helped me move closer to the type of vivid storytelling to which I aspired. I was lucky to have wonderful research assistants, who dove in and quickly familiarized themselves with research sites that often proved difficult to access: Liza Anulao, Samantha Ghai, and Megan Macomber, I am grateful for your tenacious and meticulous work.

There are many others who contributed to the ideas and logistics that went into this book that I would like to acknowledge: Lisa

Aronson, Tina Beyene, Marisol Díaz, Kolleen Duley, Kate Graney, Wendy Allison Lee, Pushkala Prasad, Rik Scarce, Rebecca Herzig, and the anonymous reviewers who read my submitted manuscript. I especially acknowledge Kimberly Guinta, my editor at Rutgers University Press, whose supportive, forthright style of communication made for a very smooth, not to mention swift, process. I am also appreciative of Terri Gomez and Carole Browner, who offered me support and guidance during the difficult job transition and relocation that happened in the midst of my writing this book; my fellow conference presentation collaborator Rajani Bhatia, who helped keep me motivated to share and present my ideas publicly; and Kavita Philip and Banu Subramaniam, who continue to influence my intellectual development. Finally, the book in its current form would not have been possible without generous funding from Skidmore College for various production aspects of the book.

My family, friends, and community were perhaps most instrumental in keeping me going through this entire process. A special thanks to my mom, who provided endless loving support. Catron Booker, Rebecca Louisell, Brinda Sarathy, Samantha Siegeler, Daphne Taylor-García, and Dominique Vuvan: you all also helped boost me up during this process, and I offer my utmost appreciation.

Notes

Introduction

1. The first of two sets were mailed out to several East Coast news locations on September 18, 2001, and the second set was postmarked October 9, 2001, to Democratic senators Tom Daschle and Patrick Leahy in Washington, DC. Over the course of several weeks, these letters caused five deaths and an additional seventeen injuries via inhalation and cutaneous anthrax in news media employees, postal workers, and others who came into contact with either the letters or the facilities they passed through.

2. Speculation about Al Qaeda drew on the organization's previously stated interest in acquiring and using anthrax against the United States. Additional reasons included the proximity of the anthrax mailings to the September 11 attacks on the World Trade Center and the Pentagon; the "Allah is Great" scribbled on some of the letters; the fact that the path of some of the 9/11 hijackers in Florida coincided with the Florida anthrax cases; and inconclusive reports that the hijackers had sought out crop dusters (to disperse biological agent spores) (Carus 2001; Center for Counterproliferation Research 2002; Rubin, Linderman, and Osterweis 2002; Thompson 2003). Speculation about Iraqi weapons capabilities and Saddam Hussein's intent to attack the United States was based on the existence of an Iraqi program active from 1985 to when it was destroyed by UN inspectors in the mid-1990s. Additional reasons included Iraq admitting to possession of a substantial amount of anthrax (along with botulinum toxin) in the 1980s (ironically, purchased from the U.S. company ATCC); and Iraq in the

late 1980s seeking out the Ames strain of anthrax (implicated in the 2001 mailings)—although it was never confirmed that Iraq acquired it (Center for Counterproliferation Research 2002; L. Cole 2003; Guillemin 2005a; O'Toole 2001; Thompson 2003).

3. The US Army Medical Research Institute for Infectious Diseases (USAMRIID), located in Fort Detrick, Maryland, was founded for the express purpose of biological defense research after the United States terminated its offensive biological weapons program in 1969. The Ames strain was originally developed in the early 1980s.

4. In 2005 the FBI had determined, through DNA sequencing and particle size testing, that the anthrax used in the mailings belonged to a batch housed in USAMRIID. Ivins was the scientist who had sole access to the flask, and the FBI began investigating him in late 2006, planning to indict him on charges related to using a weapon of mass destruction—but the case ended when Ivins committed suicide in August 2008 (Department of Justice 2010). After the formal investigation closed, the National Research Council commissioned a group of scientists to examine the evidence, namely, genetic similarities linking Ivins's batch to the anthrax used in the mailings; the commission did confirm the genetic similarities, but also pointed out that this did not definitively link them, as the two different batches could have evolved with similar genetic profiles independently (National Research Council 2011, 20). With no further developments, Ivins remains the prime suspect.

5. Seth Carus conducted a comprehensive study of bioterrorism, which he defined as "instances in which a non-state actor . . . allegedly used, threatened to use, acquired, attempted to acquire, or even expressed an interest in biological agents [infectious substances and their toxins that can be used as weapons]." He found that, of the thirty-some odd cases in the entire twentieth century targeting the United States or its inhabitants where the perpetrator was actually identified, about 97 percent were perpetrated by groups based in the United States, and the vast majority—over 80 percent—were perpetrated by white males (Carus 2001).

6. The U.S. Army devised, for example, the Human Terrain System in 2006 to recruit anthropologists and other social scientists to provide

military commanders and staff with cultural knowledge of Afghanistan and Iraq. It lasted until 2014.

7. Until the anthrax used in the mailings was definitively traced back to USAMRIID in 2005, the FBI had continued to consider Al Qaeda and Iraq, as well as pursuing other leads outside of U.S. biodefense, such as persons in geographical proximity to the anthrax mailing sites (Department of Justice 2010).

8. Although not my focus here, anthrax profilers also directly bolstered white masculinity. See D'Arcangelis 2015 for details on how, in response to the 2001 anthrax mailings, the FBI investigators shored up "white scientific masculinity" by quashing its connection to bioterrorism—through a series of profiling practices that anomalized the perpetrator as a "loner," an "amateur scientist," and finally a mad scientist.

9. U.S. geopolitical power advanced substantially with the end of World War II. The subsequent conflict between the capitalist-aligned United States and Western Europe on the one hand and the socialist-aligned USSR and Eastern European states on the other—dubbed the Cold War—was not one between two equal sides, as the latter side possessed significantly less economic and productive power (Prashad 2007, 7–8).

10. Foregoing a detailed discussion of the contentious definitions of "terrorism" and "terrorists," suffice it to say that the United States applies the terms to nonstate actors who utilize violence that threatens the government in some way.

11. The Anti-Terrorism and Effective Death Penalty Act of 1996 (AEDPA) is considered the predecessor of the PATRIOT Act, whose targeting of Arabs and Muslims I detail in chapter 1.

12. Racialization has been used as a tool of social control since the founding of the United States. European settler-colonists used racial discourses of threat and incivility against Native Americans to rationalize land dispossession; settlers portrayed African Americans as backward during the country's founding in order to justify enslaving them for white profit. White groups have also used these discursive practices against immigrant groups, notably Asian Americans and Latin Americans, depicting them as dirty and diseased in order to

rationalize their subordination. I detail the racialization of Arabs and Muslims in chapter 1.

13. I use the slash to denote their conflation in dominant discourse; more on this historical process in chapter 1.

14. The invasion of Afghanistan on October 7, 2001, was an unprecedented—and questionable on legal, moral, and practical terms—move to attack a state (Afghanistan) based on the presence of a nonstate actor (Al Qaeda) within that state. The invasion of Iraq on March 20, 2003, was without any provocation and was rationalized as preemptive action. Both military operations lasted well into the following decade, and occupation by the United States (and its allies) long after that.

15. Lynn Itagaki has elaborated on the relationship between post–Cold War era national security narratives and domestic politics—namely, post–civil rights era discourses of racial equality help produce the illusion of a United States as exceptionally democratic and civil rights affirming, which then in turn pushed from view the era's economic recession and its intensified wealth gap and interracial strife (e.g., the LA riots) (Itagaki 2016).

16. In the first month of the anthrax investigation (October 2001), the Access World News database (comprising 624 U.S. sources) contained over 8,000 articles on bioterrorism, a sharp increase in comparison to previous months, which never contained more than double-digit numbers.

17. Other news outlets followed. For example, the *Los Angeles Times* inaugurated its own "Biological Threat" section in late 2001; it was added to a series titled "The U.S. Strikes Back" (begun in the 1998 Clinton era focus on terrorism).

18. I provide more examples of this bioterror landscape in *The Bio Scare: Anthrax, Smallpox, SARS, Flu and Post-9/11 U.S. Empire* (D'Arcangelis 2009).

19. It put out its first issue in March 2003. In 2015, it was renamed *Health Security*.

20. "Technoscience" is the preferred term of science studies scholars and highlights, as I wish to, the messy, overlapping totality of science (typically associated with basic knowledge), technology (typically associated with applied knowledge), and the social.

21. In addition to the United States, Britain, Germany, Canada, Japan, and the former Soviet Union had these programs.

22. The toxic products of germs were also made into biological weapons. Biological weapons were developed to harm humans and their immediate environment (agriculture, water supply, animal farms). Germs and toxins can be weaponized via a number of means, i.e., cultivated by size, potency, dispersibility, etc., to be effective for intentionally spreading death and destruction. Examples include *Bacillus anthracis* bacterium (anthrax), variola virus (smallpox), and *Clostridium botulinum* toxin (botulism).

23. Aum Shinrikyo was a Japanese religious cult that waged several failed attempts in the early 1990s to deploy anthrax and botulinum toxin; the cult had successfully waged a chemical weapons attack with sarin in 1995 on a Tokyo subway that killed twelve and injured several hundred (Clinton 1999).

24. U.S. national security concerns were informed by testimony, not entirely verified, from Russian defectors that the former USSR had maintained an offensive program manufacturing biological agents and experimenting with smallpox from the 1970s until the early 1990s (Ahuja 2016; Hart 2006).

25. U.S. attention to biological weapons programs internationally was selective, as there were other countries known to have had strong bioweapons programs well into the late twentieth, if not the early twenty-first, century that the United States largely ignored as a threat—namely, South Africa and Israel (Barnaby 2000; Dando 2006).

26. In contrast, earlier, successful, attempts garnered no such attention, for example, a biological attack in 1984 by the Rajneeshee cult that had resulted in the contraction of salmonella by over 700 people in Oregon.

27. Lakoff (2008b) traces the history of "preparedness" to Cold War era civil defense, which began with a focus on nuclear emergencies but eventually expanded to other forms of emergency (409).

28. I draw on Robin DiAngelo's (2018) notion of "fragility" to underscore the way that dominant entities view themselves as vulnerable, despite their relative power and privilege. DiAngelo coined the phrase "white fragility" to denote white expectations for racial comfort and a lowered ability to tolerate racial stress, despite the fact that whites live in an

environment of racial protection; DiAngelo highlights how white fragility triggers defensive actions (e.g., anger, dismissals) that function to reinstate white racial equilibrium. I use the phrase "U.S. fragility" to denote the affective dimension of U.S. aggression: a sense of vulnerability triggers defensive actions to maintain global hegemony. In subsequent chapters I further break down this imagined U.S. nation as white, patriarchal, and fundamentally tied to the interests of dominant groups within the United States.

29. Cultural studies scholar Neel Ahuja elaborates on how the United States appropriates the figure of the nonimmune Indian to render itself as vulnerable to biological weapons (2016, 143).

30. DHS incorporated twenty-two agencies, domains from immigration to agriculture to health to coast guard, and hazards from terrorist attacks to epidemics and hurricanes. From the Department of Health and Human Services specifically, DHS absorbed the National Disaster Medical System and the Strategic National Stockpile (a repository of vaccines and other countermeasures).

31. U.S. public health officials have also historically treated black populations as diseased, constructing them as having an innately heightened sexual appetite, and more prone to spreading venereal disease than whites (Jones 1993).

32. SARS, for example, was articulated in U.S. public health and news media as the product of China's animal husbandry (too great a mix of animals), food markets (eating "exotic" animals), and consumption practices (butchering practices in view of consumers). See D'Arcangelis 2008 for details.

33. According to one policy analyst, quoted in the *Washington Post*, "You can't rule out that this [SARS] is a weapon" (McCombs 2003); and according to another, quoted in the *New York Times*, "It's [i.e., SARS] a very unusual outbreak . . . it's hard to say whether it's deliberate or natural" (Broad 2003).

34. According to one journalist from the *Washington Post*, biowarfare experts were avidly discussing "the specter of terrorists hiring scientists who can insert a toxin into, say, a bioengineered SARS virus, which would then be as contagious as severe acute respiratory syndrome and as fatal as the toxin inside it" (Mintz 2004). Disease specialists pointed

to the "darker side to the relation between naturally emerging infections and bioterrorism. . . . Whereas clinicians and policy makers view diseases like SARS as public-health threats, terrorists could see them as weapons of opportunity" (Weber et al. 2004).

35. The influence of national security discourse over the theory and practice of disease control reached a point where naturally arising diseases were being thought of in relation to bioterrorism. An article in the *American Journal of Public Health* even offered a new term when it suggested that government leaders should "discuss and develop effective detection and response strategies for bioterrorist and *nonterrorist* occurrences of infectious disease" (Martin 2004; emphasis added).

36. Postcolonial theorist Ania Loomba demarcates European imperialism from earlier forms (such as the Roman and Mongol empires) through the way it restructured the economies of colonized countries and established a continual flow of human and other resources between colonized and colonial countries, feeding the growth of European capitalism and industry (2015, 21).

37. Here the Global North signals the United States, its allies, and the corporate entities (i.e., biotechnology multinationals) that reign in the post–Cold War neoliberal world era of U.S.-led globalization. Transnational feminist scholar Inderpal Grewal (2005) has described in *Transnational America: Feminisms, Diasporas, Neoliberalisms* how, at the turn of the twenty-first century, the United States remained "a hegemon" whose "source of power was its ability to generate forms of regulation across particular connectivities that emerged as independent as well as to recuperate the historicized inequalities generated by earlier phases of imperialism" (21). In other words, twenty-first-century U.S. empire both accommodated and incorporated new centers of power during the 1990s.

38. Anthropologist George Marcus provides an excellent description of the difference between multisited studies and single-site-focused studies in "Ethnography in/of the World System: The Emergence of Multi-sited Ethnography" (1995). He highlights the way in which Donna Haraway ("A Cyborg Manifesto") and Emily Martin (*Flexible Bodies*), engaging science and technology studies and sociocultural

studies of medicine respectively, trace cultural formations such as discourses about the immune system *across multiple sites*—so that the researcher can learn about world systems such as capitalism and imperialism. In contrast, single-sited studies produce intensively focused characterizations of particular institutional sites—I offer as an example the work of anthropologists and sociologists of science in Lakoff and Collier's *Biosecurity Interventions* (2008), who elaborate on new "biosecurity" institutional formations in *single sites* such as the Department of Homeland Security or the Department of Health and Human Services.

39. See Nikolas Rose's *Powers of Freedom* for a genealogy of Foucault's evolving conceptions of governmentality (1999). Some of Foucault's works where he applies the concept to a variety of settings include *Discipline and Punish* (1977), *The History of Sexuality: An Introduction* (1978), and *The Care of the Self* (1986).

40. For more on the relationship between media and discourse, and the constraints this relationship places on the way journalists assemble their narratives (and the way readers consume these narratives), see Hall's "Encoding/Decoding" (1980) and feminist theorist and cultural critic bell hooks' *Reel to Real* (1996).

41. I included the *Washington Post*, which is ranked at number seven, over the ones ranked above it (the *New York Post* and the *Daily News*) due to its emphasis on national security–related coverage.

42. Access World News is a "fully searchable Web-based resource [that] features the vast majority of U.S. newspapers by circulation, along with almost one thousand hard-to-find local and regional titles, the majority of which are unavailable elsewhere" (NewsBank 2008).

43. The following exemplify the works I am in conversation with from feminist science studies: Banu Subramaniam's (2001) work on gendered and raced threat discourse in conservation biology (namely, xenophobic invasion metaphors) examines the way that U.S. nationalist discourse in the late twentieth century circulated across realms, shaping scientific ideas about "alien" plant and nonhuman animal species; Kavita Philip (2004) has detailed the history of nineteenth-century science in colonial India, unpacking how the British Empire's colonial ideologies and institutions shaped the character of science in both local

and global ways, constituting a pivotal developmental point in the legacy of science and its imbrication with the subjugation of colonized peoples.

44. The phrase "biological threats" first appeared (infrequently) in the mid-1980s to signify a whole host of meanings, from threats to ecology by new species of fish, oil spills, and population growth, to the threat to humans and other organisms by disease migration, toxic chemicals, and biology lab accidents. At the close of the twentieth century, usage of the term narrowed significantly—mainly denoting disease-causing pathogens. In the context of the bioterror imaginary, the term would increasingly represent a securitized conception of infectious disease emergence—caused by naturally arising disease as well as human-made biological weaponry. (I conducted this genealogy by searching the following online databases: ProQuest News, Access World News, ProQuest Congressional [formerly LexisNexis Congressional], HighWire Press, and PubMed.)

1. The Making of the Technoscientific Other

1. Even though many among the military believed that biological weapons could be as effective and lethal as nuclear weapons, they still supported the ban, believing that without the ban other countries might be able to keep pace with U.S. biological weapons development (thus eliminating any U.S. strategic advantage) (Kelle, Nixdorf, and Dando 2012, 139).

2. When I refer to "Western," I am using Stuart Hall's conception: it denotes a complex of ideas about society and progress that originated in Western Europe during the sixteenth century but eventually migrated beyond this geography—it is primarily a historical construct denoting societies that have industrialized and are secular; "Western" applies, for example, to Japan but not to Eastern Europe (Hall 1992, 276–277).

3. Since their arrival in the United States, Arabs had been ambivalently treated as white, a tenuous status that meant they were not always guaranteed white privileges such as citizenship (Majaj 1999; Naber 2000; Samhan 1999).

4. Prior to its entrenchment in U.S. society, race originated as a structuring principle in European colonialism in the fifteenth century, maintaining hierarchies of rule not only between nations but also within local contexts. Racial formation has been a worldwide process wherein dominant groups have used race to create hierarchies of rule, what Asian and Asian American studies scholar Shu-mei Shih calls the "worldliness of race" (2008). Racial formation has also proven quite malleable, incorporating both cultural and biological syntaxes of difference to racialize its subjects. Thus, even as the racialization of Arabs relied distinctly on cultural essentialisms, at times racialization relied on notions of biological difference—e.g., color and other sloppy phenotypic categories (Cainkar 2008).

5. The rise of a politically conservative Islam was due in no small part to U.S. intervention. During the late Cold War period the CIA conducted a worldwide campaign to recruit, train, and support militant, politically conservative Islamic guerrillas as soldiers to fight as proxies in the U.S. war against the Soviet Union (Mamdani 2004a).

6. The Byzantines viewed Arabs as primitive and sexually immoral, and Islam, when it arose in the region, as dark and evil. These views structured the views of Western Europe and later European colonists in the Americas (Naber 2000).

7. A foreign national is a citizen of a foreign country who does not have permanent residency in the United States and is often on a student or visiting visa. The PATRIOT Act altered the rights guaranteed foreign nationals and other noncitizens suspected of terrorism by granting government authority to indefinitely detain them without process or appeal. The application of "terrorism" charges, moreover, was broadened to include providing "material support to terrorist organizations" and "mass destruction." The act also authorized, in the name of counterterrorism, enhanced government surveillance capabilities via phone and Internet, and delayed-warrant searches.

8. Paul Amar (2011) has detailed mainstream U.S. constructions of Arab/ Muslim masculinity, and how they prop up the logics of what he calls "terrorology industries" (38).

9. In addition, federal agents conducted "voluntary" interviews of thousands of men ages eighteen to thirty-three who entered the United

States after January 2000 and were on nonimmigrant visas—most were Muslim and/or Arab and from countries where Al Qaeda was thought to have a presence; and the State Department implemented new visa screening procedures targeting men ages sixteen to forty-five from Arab and/or Muslim countries with twenty-day waiting periods and extra security checks.

10. These were on the State Department's list of state sponsors of terrorism in 2001; North Korea and Cuba were also on this list, but not targeted for special registration (Department of State 2002).

11. The only country on this list not Arab or Muslim was North Korea; the list includes Iran, Iraq, Libya, Sudan, Syria; Afghanistan, Algeria, Bahrain, Eritrea, Lebanon, Morocco, Oman, Qatar, Somalia, Tunisia, United Arab Emirates, Yemen; Pakistan, Saudi Arabia; Bangladesh, Egypt, Indonesia, Jordan, and Kuwait.

12. Nadine Naber has discussed how class is also a factor in racializing Arabs and Muslims: that a lower class status draws out racialized tropes even more—both men as "potential terrorists" and women as passive victims (Naber 2008).

13. Much fewer were the representations that attempted to avoid the trope of Arab/Muslim cultural Otherness. Arab and Muslim American studies scholar Evelyn Alsultany (2008) describes, for example, the popular TV representations after September 11 that attempted—not entirely successfully in her view—to distance themselves from discourses of terrorism by portraying Arab and Muslim Americans as something other than terrorists (as, for example, a foreign exchange student or a handyman).

14. Rana (2013) and Puar (2007b) describe the way immigrants of color get lumped together, in effect functioning as an amorphous and broad target for state-sanctioned violence.

15. Several scholars use "Arab/Middle Eastern/Muslim" to designate this enlarged grouping. Given that even this expanded categorization does not encompass everyone caught up in this increasingly messy racialization process, I continue to use "Arab/Muslim" for the sake of simplicity.

16. See also Amar 2011; Bhattacharyya 2008; and Muscati 2002.

17. See Cameron (2009) for more on the trope of the cowardly terrorist and the figure's unacceptable, immoral behavior.

18. While some, namely, mainstream social scientists, acknowledge that such violence may be the product of political strategy—à la rational strategic choice theory—they do not negate the features of backwardness, inferiority, and irrationality in their formulation so much as reduce them to variables that can be quantified, and thus subject to calculation and control (see, for example, Caplan 2006; Crenshaw 1990; Sandler 2003).

19. I traced the origins of the term (via Nexis Uni [formerly LexisNexis Academic] and ProQuest News, as well as Google Scholar and even Google) to its first use in early October 2001, just after the anthrax attacks.

20. Smallpox, a particularly deadly disease, had long been the concern of U.S. weapons specialists. Anyone who acquired and deployed smallpox could wreak havoc and deaths in the millions, and smallpox became an even greater focal point during the war on terror. I focus on the politics around smallpox vaccination in the next chapter.

21. One text was explicit about this connection, describing the figure as "the biological equivalent of a suicide bomber" (Chase 2001).

22. As described in the introduction, while I typically focused on five of the most circulated newspapers, I also utilized a comprehensive database, Access World News, comprising 624 U.S. sources, to more generally survey topics related to bioterrorism. The "suicide infector" and its various permutations was a prominent theme articulated across many news venues.

23. In my research, the suicide infector's gender was, if specified, marked as masculine.

24. The Iraqi bioweapons program, which began development as early as 1974, and was part of Iraq's overall CBRN (chemical, biological, radiological, and nuclear weapons, which in the United States we colloquially refer to as WMDs) program, was destroyed in 1996 under the supervision of the United Nations Special Commission (UNSCOM). This was the culmination of inspections that had begun in 1991, after Iraq lost the Gulf War. Even after tensions terminated inspections at the end of 1998 (which was subsequently followed by U.S./UK bombing known as Operation Desert Fox), new inspections were eventually reinstated in 2002 with the passing of UN Resolution

1441. Iraq complied with inspections and filed a weapons report over 10,000 pages long by the end of 2002, and UNMOVIC (the successor organization to UNSCOM) found no evidence of any renewed operations (Guillemin 2005a; Nuclear Threat Initiative n.d.; UNSCOM 1999).

25. National security information is hard to come by, as it is by nature a secretive realm. My limitation to U.S. sources was also a hindrance to collecting information about the Iraqi scientists. I acquired much of my information about their involvement (as well as their biographies and their detention) via U.S. news media and academic sources, both of which drew heavily on the sparse quotes of government officials and biodefense specialists in the U.S. and international arms community. What I culled generally outlined the following: Dr. Taha had reportedly played a role in the Iraqi biological weapons program during the early part of the Iran-Iraq War, and had been the subject of UK and U.S. surveillance since the mid-1990s during that first round of UN weapons inspections. Dr. Ammash had gained attention just before the 2003 invasion of Iraq, after being spotted in a photo with other top Iraqi officials; she was reportedly suspected of playing a key role in developing the biological weapons arsenal in the period following the First Gulf War.

26. During the First Gulf War of 1991, the United States mobilized narratives of Saddam Hussein as the ultimate Arab/Muslim threat (made easier as he adopted the mantle of Islam and Arab nationalism) to justify its invasion of Iraq and the subsequent sanctions that led to unconscionable numbers of Iraqi deaths, especially children (Muscati 2002).

27. See Mamdani, "From Proxy War to Open Aggression" (2004c).

28. The dupe caricature developed in connection with Arab and Muslim women engaged in suicide bombing: they defy stereotypical notions of the passive Arab/Muslim woman, and are typically portrayed as disempowered dupes whose actions can be explained in terms of personality flaws, or as motivated by familial loyalties to their husbands and fathers—all tropes that obscure the political motives of their actions (Amireh 2011; Brunner 2007).

29. Ella Shohat (1991) overviews the way tropes of geography and femininity intertwine in service of colonialism to portray subjugated

women—from women of color in the United States to women in the Middle East—as dark and libidinous (or alternatively as virginal and tameable).

2. From Practicing Safe Science to Keeping Science out of "Dangerous Hands"

1. The term "biological agent" was actually codified in the 1989 Biological Weapons Anti-Terrorism Act with respect to biological warfare to denote "any micro-organism, virus, infectious substance producing death, disease, other biological malfunction or deterioration of resources or deleterious alterings of environment."

2. Geopolitical aims have often guided U.S. actions more than the dictates of international treaties: another prominent example was when the United States granted Japan immunity from war crimes prosecution after World War II (the latter's vicious actions toward China included an estimated 10,000 killed due to biological warfare actions and experiments) in exchange for knowledge gained from their experiments (Rosenbaum 1998).

3. See, for example, "President Bush Outlines Iraqi Threat" (Bush 2002a).

4. See Shannon Steen (2010), among others, for a discussion of the multicultural face of U.S. imperialism.

5. The inspection was organized by Rooting Out Evil, a Canadian-based coalition of international groups opposed to the development, storage, and use of weapons of mass destruction by any nation. More information on the group can be found at http://www.socialjustice.org/index .php?page=peace-justice.

6. A search of ProQuest News yielded about fifteen articles on the event. The coverage, moreover, entailed brief, superficial mentions in, for example, the *Washington Post* (Markon 2003) and the *Boston Globe* (Robertson 2003), and some coverage in progressive news outlets such as the *People's World* (Chicago, IL), *City Pulse* (Lansing, MI), and the Institute for Public Accuracy (Washington, DC).

7. King and Strauss (1990) detail the various methods of vaccine development. They note that some of the methods are less dangerous than others, as they involve using smaller amounts of pathogenic material,

such as preparation from attenuated strains or utilization of smaller protein subunits of the pathogen, but that these methods have proven fairly ineffective against pathogens that have many variants.

8. The full text of the relevant section is: "Each State Party to this Convention undertakes never in any circumstances to develop, produce, stockpile or otherwise retain: (1) Microbial or other biological agents, or toxins, whatever their origin or method of production, of types and in quantities that have no justification for prophylactic, protective or other peaceful purposes; (2) Weapons, equipment or means of delivery designed to use such agents or toxins for hostile purposes or in armed conflict." For full text of the BWC, see http://www.opbw.org/convention/conv.html.

9. S. Wright and Ketcham (1990) have suggested that a purely defensive program must limit itself to developing generic defense components such as generic therapy, detection, decontamination, and protective clothing (188).

10. Russian president Boris Yeltsin acknowledged only that there had been a lag in the Soviet Union's implementation of the BWC. Accounting for the Soviet program and its offensive capacity has in fact been quite difficult, as evidence derives from memoirs of former Soviet scientists who defected, Soviet scientific articles on defensive aspects of research on biological agents, and the declassified intelligence documents of other countries. John Hart (2006) has detailed significant evidence of Soviet offensive capabilities (from 1973 to 1992); however, as he also admits, there is no authoritative or comprehensive account of the Soviet program from either oral histories or a systematic study of primary documents. Neel Ahuja (2016), moreover, has pointed out that some of the claims of defectors have been contradicted (namely, those of Dr. Kanatjan Alibekov, who Anglicized his name to Ken Alibek, former deputy chief of research and production for the Soviet civilian program Biopreparat).

11. Although much of the biodefense industry remained in the hands of the Department of Defense, this decentralized program enlisted the Department of Justice and the Federal Emergency Management Agency (FEMA), and bolstered state and local agencies' role. It also emphasized broad technological solutions and emergency response,

establishing the National Pharmaceutical Stockpile and the Health Alert Network and Laboratory Response Network in 1999 (Guillemin 2005a; Bernstein 1987; Khan, Morse, and Lillibridge 2000).

12. Significant allocations also went to infrastructural modifications in health and science, such as improved communication systems, planning, training exercises, and high-tech detection equipment (*Budget of the U.S. Government* 2003; "HHS Fact Sheet" 2004). This included, for example, the nationwide system of air monitoring devices known as BioWatch, which was set up in thirty-one cities across the country for real-time surveillance of aerosolized pathogens (Shea and Lister 2003).

13. Specifically, for the 2004 budget, the NIH was allocated $1.6 billion (an increase of $121 million from the prior year—and significantly more than the mere $53 million allocated for 2001) for researching and developing new biodefense countermeasures ("HHS Fact Sheet" 2004).

14. See Ismail 2007 and Ridgeway 2005 for details.

15. Vaccines are not as lucrative as other pharma products because they are single-use (in fact, many vaccines are made by only a handful of companies, such as Bayer). Accordingly, the federal government initiated a slew of measures to incentivize pharmaceutical makers to participate in the biodefense market: Project BioShield allocated $5.6 billion to drug companies over ten years to produce vaccines and other drugs; this was followed by the passage of the Pandemic and All-Hazards Preparedness Act in December 2006, which introduced even more tax incentives and greater compensation to companies at earlier stages in the countermeasure development process.

16. Since 1980, review conferences have been held every five years to make sure countries are in compliance with the protocol.

17. There was also growing protest against U.S. deployment of chemical weapons and herbicides in Vietnam.

18. The section is titled "Expressing the Sense of the Senate concerning the Provision of Funding for Bioterrorism Preparedness and Response" and specifies the goal "to better prepare the United States to respond to potential bioterrorism attacks," outlining investments in expertise and resources for bioterrorism preparedness.

19. I detail this in chapter 1.

20. The 2006 map, suggestively retitled "Protection or Proliferation?," was accompanied by a document cataloguing the incidents. A link to this map and list may be found online at http://web.archive.org/web /20100716132010/http://www.sunshine-project.org/biodefense/.

21. Beyond accidents, the biological warfare industry generates other significant hazards on the domestic front, namely, through field testing of biological agents in remote locations or with simulants in populated areas—neither has always proven to be harmless (L. Cole 1997; Hersh 1968).

22. "Biosecurity" had a broader signification prior to this period: it previously related to arms control and to health and agriculture issues, such as preventing nonhuman animal diseases or agricultural pests from entry into a country (Malakoff 2004; Sunshine Project 2003).

23. The PATRIOT Act built on the Biological Weapons Anti-Terrorism Act of 1989 (BWATA), which implemented a section of the BWC that encoded the first legal establishment of criminal penalties for individuals who conduct biological weapons research unless for peaceful research and development. AEDPA had further tightened regulations on biological agents: expanding the definition to account for bioengineering capabilities, including "attempts" as punishable offenses, and establishing more enforcement and safety procedures, such as transfer rules and a list of agents maintained by HHS.

24. The State Department began its list of "state sponsors of terrorism" in 1979 with Libya, Iraq, South Yemen, and Syria, pursuant to section 6(j) of the Export Administration Act of 1979 (P.L. 96–72).

25. Chapter 1's section on Orientalism contains background on how racialization processes are the product of geopolitical imperatives: Arab regions and thus Arab nationals have been racialized as terrorist. See Lisa Lowe's "Immigration, Citizenship, Racialization," in *Immigrant Acts* (1996) for an insightful history on how U.S. geopolitics racializes nations and thus nationals who migrate to the United States (her particular focus is nations in Latin America and East Asia).

26. Postcolonial theorist Rey Chow (2002) elaborates Foucault's famous theorization of biopower, a technology that generates and optimizes life in part through oppressive mechanisms—racism being one significant manifestation—which get justified in the name of the

living. Cultural-political geographer Ben Anderson discusses post-9/11 iterations of biopower that emerged to legitimize forms of intervention to optimize "valued life" against threat: from surveillance of suspicious credit card activities to emergency planning for the aftermath of a terror attack (2012).

27. Steve Kurtz was a member of the Critical Art Ensemble, an artists' collective that develops projects to address the politics of biotechnology via books, performance art installations, and often using scientific equipment and nonpathogenic organisms (da Costa 2010).

28. The late Beatriz da Costa (2010) summarized the aftermath of Hope's tragic death: "Hope Kurtz, one of the original members of Critical Art Ensemble died in her sleep of heart failure on the night of May 11th this year. Her husband and university professor Steve Kurtz called 911 after waking up next to his dead wife. The local police came to his house, searched the surroundings, and confiscated Hope Kurtz' body in order to determine the cause of her death. (After it had been cleared by the Erie County Medical Examiner, the FBI seized the body again and returned it a week later.) During their visit, the police took note of Critical Art Ensemble's mobile DNA extraction lab. The following day, Steve Kurtz was detained by members of the Federal Bureau of Investigation and representatives of the Special Task Force on Terrorism."

29. Language from section 817, "Expansion of the Biological Weapons Statute."

30. In section 351A, "Title II—Enhancing Controls on Dangerous Biological Agents and Toxins."

31. Concern with maintaining the ideal of openness in publishing and collaboration echoed the controversies of sixties-era scientist activism: alongside protestations of the role of biology in warfare, there had also been an uproar among scientists about muzzling academic and scientific freedom—after it was revealed in 1965 that scientists at the University of Pennsylvania had been working in secret on chemical and biological warfare projects under government contract and were not free to publish their research results (Clarke 1968).

32. Not everyone in the U.S. biodefense community was as sanguine about such risky research; see Jonathan B. Tucker's "Avoiding the Biological Security Dilemma" (2006).

33. Carol Cohn's (1987) pivotal work on nuclear defense intellectuals demonstrated how their rhetorics—of domestication, for instance— served to reframe the uncontrollable forces of nuclear science and destruction as, in fact, controllable.

34. I first deployed the concept "white scientific masculinity" to note its function during the anthrax investigation, namely, bolstering the U.S. biodefense industry. The anthrax investigation had produced a significant rupture in the status of white scientific masculinity—the FBI naming of Bruce Ivins (a white male U.S. biodefense scientist) as the perpetrator of the anthrax mailings. I showed that this disruption did little to alter the narrative of protective white scientific masculinity juxtaposed against the Arab/Muslim bioterrorist Other (D'Arcangelis 2015).

35. Smallpox became the first infectious disease to be completely eradicated from nature in 1979, the culmination of public health efforts across the globe. These efforts were ostensibly led by the World Health Assembly (the policy-setting body of the United Nations World Health Association), and eradication was officially certified on May 8, 1980, by the World Health Assembly. (The only other disease to be eradicated as of this writing is rinderpest, a virus infecting cattle.)

36. Additional concern stemmed from the fact that smallpox had been the object of research and development in several bioweapons programs since World War II (Ahuja 2016).

37. Russia later moved the stocks to Vektor laboratories in Novosibirsk, Russia.

38. Even though, as mentioned earlier, Iraqi weapons programs had been subject to inspections after the First Gulf War, in the mid-1990s segments of the U.S. national security and public health spheres were concerned about possible Soviet-Iraqi links or possible weaponization of variola retained from a natural outbreak in Iraq during the early 1970s (Nuclear Threat Initiative 2001, 2002).

39. Both the sequencing of the entire variola genome and the preservation of fragments of variola DNA sequences could be used in place of viral stocks for confirmation of the identity of a smallpox-like virus and other diagnostic activities.

40. Critical voices had included the Federation of American Scientists, the Society for Social Responsibility in Science, and to a lesser degree the

American Association for the Advancement of Science and the American National Academy of Sciences.

41. His was an early voice highlighting the importance of biosafety, which would gain traction in 2007 with media and government attention to accidents at Texas A&M University.

3. Co-opting Caregiving

1. In 1979, smallpox became the first infectious disease to be completely eradicated from nature, the culmination of public health efforts across the globe. It exists only in viral form.

2. The controversy over the Anthrax Vaccine Immunization Program illustrates how mandated vaccination and informed consent has applied to the military (Black 2007).

3. Polls of physicians, nurses, and other health care personnel in November and December 2002 around the time of President Bush's announcement of the NSVP indicated a willingness to receive smallpox vaccination at a rate of 61 percent in one study and 73 percent in another (Everett et al. 2003; Yih et al. 2003).

4. Feminist historians have highlighted the long history of progressive nurse activism, from agitating against low pay and poor working conditions to racial exclusion within nursing's ranks (Hine 1989; Reverby 1987).

5. The Johns Hopkins Center for Civilian Biodefense Strategies, one of many civilian biodefense centers that had sprung up in the late 1990s, had commissioned the exercise. See "From Population to Vital System: National Security and the Changing Object of Public Health" in *Biosecurity Interventions* (Lakoff 2008a) for more details on the exercise's genesis and participants.

6. Other studies showed much longer-lasting immunity—30 years, for instance (J. Cohen 2001).

7. The simulated smallpox attack entailed the infection of 1,000 people at each of three separate locations nationally and presumed an amount of smallpox vaccine based on what was actually available at the time.

8. Some researchers posited lower transmission values (due to, for example, the less crowded living conditions and increased sanitation in comparison with earlier eras) (Enserink 2002; Guillemin 2005a).

9. Pundits used additional terms to signify the dual benefit or synergy thesis, including "double merit" and "dual use," although the latter term overlaps with the negatively valenced concept denoting biological research that can potentially be misused.

10. Historian of medicine Nicholas B. King (2004) has traced the institutional overlap between health and national defense to earlier connections that were forged, in fact, strategically from the health field. As King notes, during the early 1950s, CDC chief epidemiologist Alexander Langmuir capitalized on the Cold War anxiety about biological warfare to channel defense funds into laboratory investigation of infectious disease and to create the Epidemic Intelligence Service. Many subsequent public health leaders made the same pitch; for example, Donald Henderson, who directed the international effort to eradicate smallpox, frequently advocated for improvements to health infrastructure, pulling resources to infectious disease control for both the threat of biological weapons and natural epidemics.

11. Some select populations continued to be vaccinated, such as military members (Grabenstein et al. 2006).

12. It is beyond the scope of this discourse analysis to engage in depth with reception theory and its exploration of how readers receive and interpret the meanings of these photos and texts. Feminist theorist and cultural critic bell hooks describes the ways that "[movie] audiences are clearly not passive and are able to pick and choose [messages]," but that "it is simultaneously true that there are certain 'received' messages that are rarely mediated by the will of the audience" (1996, 3).

13. I collected and analyzed photos from three of the top four most widely read newspapers with photo indexes: *USA Today*, the *New York Times*, and the *Los Angeles Times*. The two 1947 photos I focus on in figure 4 appeared in those three venues alone by my count eight times in the case of the first photo and three times in the case of the second between October 2001 (when attention to smallpox spiked) and October 2003 (a few months after the NSVP's decline). Note that I reproduce in print here a *New York Times* article because only the *New York Times* makes the images electronically viewable (versus hard copy on microfilm / microfiche).

14. Science and medicine were imbricated in the discursive milieu of their contexts, in this case the dualistic separation of dominant and subordinate populations as clean/dirty, civilized/uncivilized, rational/uncontrolled, and the like (Lupton 1995; Marks 1997; Vaughan 1991).

15. For example, at the turn of the twentieth century the Public Health Service in San Francisco conflated the Chinese race and the spread of bubonic plague in their health policies along the Pacific Rim. This led to many discriminatory practices, such as refusing Chinese entry into the United States and targeting Chinese already in the country with undue quarantine and sanitation. The specter of Chinese contagion was so strong that after one Chinese man was found dead of what was believed to be plague, San Francisco public health authorities quarantined and disinfected the entire San Francisco Chinatown, and removed whites from the area (Shah 2001).

16. The number of U.S. Chinatown businesses across the nation dropped 50 percent during the scare (Tung 2003); Chinese American children were taunted on the playground (Newman and Zhao 2003).

17. This trope stems from Cold War constructions of an inscrutable, secretive China, a representation that served to uphold the global position of the United States (Kim 2010).

18. Photos of diseased bodies of color appeared sporadically. The photos in general that emerged within and alongside articles covering post-9/11 smallpox preparedness coalesced around a few prominent themes: experts involved in various aspects of vaccination implementation (photos of decision-making panels, head shots of individual doctors and health officials); individuals receiving the vaccine (mostly government leaders, health workers, military, and civilians); and research and vaccine information as well as recipients showing symptoms of adverse reactions.

19. Whereas Chinese and other Asian immigrants have always been negatively racialized, some European populations such as the Irish have had a shifting relationship to whiteness and the privileges that go along with it. Catherine Eagan (2003) and Peter D. O'Neill (2017) detail the complexities of Irish American racialization in the nineteenth and early twentieth centuries: Irish were always categorized as "white" in legal status, but culturally racialized as inferior—mediated

in large degree by Anglos' concern about Irish Catholicism and nationalism. Over the course of the twentieth century Irish immigrants, like Eastern European and Southern European immigrants, categorically shifted to fully white status.

20. Mallon was deemed a public menace and isolated for more than twenty years on New York's North Brother Island (Leavitt 1996).

21. Three deaths (two in the civilian program, one in the military program) that were thought to be vaccine-related occurred in late March 2003, and there had been additional "adverse events" that ranged from minimal to life-threatening, such as encephalitis. For details on the adverse events that occurred during the NSVP, see Kuhles and Ackman 2003. A later study of civilian smallpox vaccination from January to October 2003 would reveal an adverse event rate of 2.17 percent (Casey, Iskander, and Roper 2005), rather high among vaccine-related complications; in comparison, routine immunizations were at .0031 percent for influenza and .0163 percent for measles, mumps, and rubella (MMR).

22. It is beyond my scope to discuss the anti-vaccination movement and its opposition to government-mandated vaccination or its claims about the negative health effects of vaccines in general.

23. My focus is on the way the image depicts whiteness, regardless of the actual race/ethnicity of the individuals in the image.

24. It is interesting to note the post-9/11 iteration of the woman-as-caretaker-of-the-nation trope. Inderpal Grewal (2006) has described the post-9/11 "security mom," the construct of a (typically middle-class and white) wife and mother who acts as an accomplice to the neoliberal U.S. security state, readying herself to protect her family and children from various security threats. With respect to the militarized health context of biodefense specifically, news articles described women stocking gas masks and other bio-preparedness items in their purses and warning their children about the mail: "Women are taking their little black Prada techno-nylon bags and slipping in gas masks for the couple [her and her husband], Cipro, a flashlight, a silicone gel tube—you shmear the silicone on your skin so hopefully it doesn't absorb the spores as fast" (Dowd 2001); "You've got housewives in rural Kentucky telling their kids to be careful with the mail" (Boyd 2001).

25. Aimee Carrillo Rowe (2004) has outlined many of the historical and post-9/11 discourses that use white women as a symbol of fragility, from instances of reasserting colonial domination to attempts to criminalize racialized immigrants.

26. Mass news media coverage of smallpox, which spiked during the anthrax scare in October 2001 and peaked in January 2003 as the NSVP began, frequently echoed the government's depiction of threat and rationale for vaccination. Coverage primarily consisted of possible sources of bioterrorism (typically mentioning Iraq), U.S. lack of preparedness for bioterrorism, the gory history of smallpox and its ravages, debates on the NSVP and the government's rationale, and, to a lesser extent, criticism of the program (particularly as health workers' resistance escalated). I culled these themes from *USA Today*, the *Wall Street Journal*, the *New York Times*, the *Los Angeles Times*, and the *Washington Post*.

27. See discussion in the Institute of Medicine report *The Smallpox Vaccination Program: Public Health in an Age of Terrorism* outlining studies that did show a connection between smallpox vaccine and cardiac complications as well as the viewpoint that cardiac events may simply have been missed in the earlier records (2005, 47).

28. As was the case with the image of "white" women I analyzed above, my focus is on the way the image depicts whiteness, regardless of the actual race/ethnicity of the individuals in the image.

29. In addition to the images of white girls, images of white boys and mixed-gender photos also circulated; I focus on the picture of female white children due to its feminized dimension, which particularly highlights the connotation of vulnerability. Note that images of white children were less frequent than images of white women getting vaccinated.

30. Individuals vaccinated with the vaccinia virus (which comprises the vaccine for smallpox) are infectious via direct or indirect contact (it is most transmissible within six feet) for up to twenty-one days after vaccination (more precisely, from the time of papule development, two to five days following vaccination, until the lesion has fully scabbed, fourteen to twenty-one days post vaccination) (CDC 2003c).

31. For a sustained discussion of nurses' actions in response to the NSVP, and the ways in which this pushed forward intersectional and transnational elements of nurse activism (which historically has focused on the needs of elite nurses), see "Confronting Public Health Imperialism: A Transnational Feminist Analysis of Critical Nurse Responses to the National Smallpox Vaccination Program of 2002" (D'Arcangelis 2019).

32. The other hindrance named in the report was the infrastructural barriers internal to the program, i.e., "the program schedule, which placed heavy demands on CDC and the jurisdictions" (Government Accounting Office 2003).

33. Frontline health care work is a feminized field that, unlike hospital administration and M.D. practice, is not lucrative or prestige-driven. Approximately four-fifths of frontline health care workers are female ("Workers Who Care" 2006).

34. Approximately one-third are people of color, which, as one study notes, is "in sharp contrast with many other health professions, in which workers are predominantly white and male" ("Workers Who Care" 2006).

35. The Iraq Body Count project has recorded thousands of deaths of Iraqi civilians each year since 2003.

36. The Department of Defense has recorded over 4,000 U.S. soldier deaths from the war's start to 2019—the vast majority occurring between 2003 and U.S. withdrawal in 2011 (Department of Defense n.d.).

37. See Alondra Nelson's *Body and Soul: The Black Panther Party and the Fight against Medical Discrimination* (2013) for an in-depth discussion of the party's fight for health care access and emancipation from medical apartheid as part of civil rights for the black community.

38. One well-known historical example is U.S. military medicine in the Philippines at the turn of the twentieth century, which sought to both safeguard U.S. officers' health and fashion Filipino bodies into what the United States deemed "hygienic" and "civilized" (W. Anderson 2006). See Greene et al. (2013) for an overall genealogy of colonial medicine.

4. Preparedness Migrates

1. The precise conditions that enable this pandemic adaptation of avian flu are still under intense debate, but are believed to occur primarily via two mechanisms. In the first, relatively rare case, avian influenza viruses gradually adapt through repeated contact with humans to the point where they can jump the so-called species barrier and acquire the ability to infect humans. The second mechanism occurs through a process known as reassortment, wherein two different strains (e.g., human and avian flu strains) infect the same cell mix (e.g., human or pig) to create a new strain—in this case a newly infectious human flu strain (Khaliq et al. 2016; Schrijver and Koch 2005).

2. The 1918 "Spanish flu" killed approximately fifty million globally and 675,000 in the United States; the 1957 "Asian flu" killed approximately two million globally and 70,000 in the United States; and the 1968 "Hong Kong flu" killed approximately one million globally and 34,000 in the United States (Institute of Medicine 2008).

3. Examples of avian flu outbreaks in humans include the following: the H7N7 subtype in England in 1996; in Hong Kong, H5N1 in 1997, H9N2 in 1999, and again H5N1 in 2003; H7N7 in the Netherlands in 2003; H5N1 in multiple Southeast Asian regions in 2004; in North America, H7N2 in 2003, and in 2004 H7N3, H7N2, and H5N2; multiple outbreaks in African and European countries in 2006 (CDC n.d.d; Schrijver and Koch 2005; Wiwanitkit 2008). In contrast, between 1959 and 1996, there were only three such cases, two of which were related to lab accidents (Schrijver and Koch 2005).

4. Human-to-human transmission has occurred primarily among the blood relatives acting as primary caregivers to infected people (and severe disease has not occurred for those infected in this way) (CDC n.d.a).

5. These U.S. models put the death toll in the United States alone at 1.9 million over several months or possibly a year (Department of Health and Human Services 2005). In contrast, World Health Organization projections put the number at an estimated several million deaths worldwide (WHO 2005d).

6. The President's National Strategy for Pandemic Influenza (Homeland Security Council 2005) released on November 1; the Department of Health and Human Services' (HHS) Pandemic Influenza Plan released on November 2; the White House's National Strategy for Pandemic Influenza Implementation Plan released on May 3, 2006 (Homeland Security Council 2006); and finally, the Department of Defense's Implementation Plan for Pandemic Influenza released in August 2006. The president's plan, for example, would outline international health surveillance and containment efforts; medical stockpiles; the domestic capacity to produce emergency supplies of pandemic vaccine and antiviral medications; and preparedness at all levels of government, all funded by a whopping $7.1 billion emergency budget supplemental request (Department of State 2005).

7. See the Center for Biosecurity of UPMC (renamed in 2013 the UPMC Center for Health Security) discussion on the questionable efficacy of these methods; for example, "There is no evidence that this type of [large-scale geographic] quarantine would slow the spread of flu, but it could have severe adverse consequences" (Center for Biosecurity of UPMC 2006). Yet, the National Strategy for Pandemic Influenza Implementation Plan focused on geographic quarantine: "the isolation, by force if necessary, of localities with documented disease transmission from localities still free of infection" (Homeland Security Council 2006); the document also delineated the following border control methods: "targeted traveler restrictions to help contain the pandemic at its source, and implementation of layered, risk-based measures, including pre-departure, en route, and arrival screening and/or quarantine," elaborating that they "may be effective in delaying the onset of a pandemic in the United States and can help minimize the risk of infection among travelers coming to the United States." The role of the military was outlined in the Department of Defense's Implementation Plan for Pandemic Influenza: on the authority of the president it would "provide support to civil authorities in the event of a civil disturbance" or assist civil authorities in "isolating and/or quarantining groups of people in order to minimize the spread of disease during an influenza pandemic" (2006).

8. Subsequent years have hovered around thirty deaths in total over a dozen countries (WHO 2013, 2015).

9. H5N1 and other highly pathogenic avian influenza (HPAI) strains have since been found among birds in the United States in late 2014 (CDC n.d.b).

10. See historian of medicine Nicholas B. King's work (2002) for a comprehensive genealogy.

11. They pointed to a variety of recent historical changes to explain the perceived rise: increasing life spans, mass production, technical sophistication in food processing, antibiotics, ecosystem disruption, intensification and monoculture in farming, international travel and commerce, microbial adaptation and change, the breakdown of public health measures, and new invasive medical procedures (Garrett 1995; Gibbs 2005; Lashley and Durham 2002; Lederberg, Shope, and Oaks 1992; Shope and Evans 1993).

12. Lorna Weir and Eric Mykhalovskiy (2010) detail, from the mid-1990s through the mid-2000s, the "internationalization" of the emerging infectious disease concept, namely, its uptake in the WHO and reformulation as a matter of global health. A U.S.-Canadian alliance met with the WHO in 1994, and again in 1995, to present its emerging diseases worldview and to pressure the WHO to adopt its priorities. As a result the WHO revamped its focus on communicable diseases, elevated emerging diseases in particular, and worked toward a more coordinated global vigilance apparatus, culminating in 2005 with the revised International Health Regulations that expanded WHO authority to regulate communicable disease threats. Moreover, cultural studies scholar Priscilla Wald has highlighted the uptake of an "outbreak narrative" wherein global interconnection is seen as the source of increased outbreaks but also of the global cooperation, particularly through scientific collaboration, paramount to containing these outbreaks (Wald 2008, 2).

13. I use the term "Global South" because I believe it best captures the geopolitical processes wherein "South" regions have been socially and economically disempowered by long-standing systems of Western colonialism as well as more recent economic systems of global capitalism. I prefer this term over "Third World," Cold War–derived

terminology that in the post–Cold War era often reduces the global divide to poverty disparities alone; or "developing countries," which no longer connotes the lessons of dependency theory (that Western powers underdeveloped "developing countries").

14. While the reasons for the rapid containment of SARS were complex (not the least of which were the characteristics of SARS itself, including its low transmissibility and the timing of symptom onset, namely, prior to peak infectivity, greatly increasing the likelihood of identification before the host spreads the virus to others), many credit the international system as paramount in that it enabled the sharing of information on diagnosis and treatment (Heymann and Rodier 2004).

15. During the SARS outbreak, the WHO utilized GOARN, established in 1997, to implement its enlarged data acquisition capacity (to include data from nonstate sources) and to facilitate international cooperation for researching the SARS virus, sharing information, and publicizing the spreading epidemic. Beginning in mid-March 2003, eleven laboratories in ten countries had collaborated to identify the SARS pathogenic agent (WHO 2003a), and clinicians shared information and experiences on the diagnosis and treatment of SARS (Institute of Medicine 2004; WHO 2003b).

16. By some accounts it was the laboratory at Hong Kong University in collaboration with Guangdong scientists that first made the identification (Lee and Warner 2008; Sung and Cheung 2003).

17. From 1851 to 1938, twelve European nations held fourteen International Sanitary Conferences, dedicated to standardizing quarantine regulations internationally; the Office Internationale d'Hygiène Publique, founded in Paris in 1907, was also important in collecting and disseminating disease information. In 1902, the United States set up the International Sanitary Office of the American Republics; and the Rockefeller Foundation International Health Division also played a key role in international health (Brown, Cueto, and Fee 2006; N. King 2002).

18. Following the advent of the war on terror, the Global North pushed the WHO to include event monitoring of CBRN (chemical, biological, radiological, and nuclear weapons, which in the United States we colloquially refer to as WMDs) under its purview in the revision of the

International Health Regulations. The ensuing resistance from WHO members in the Global South (reflected in the Montevideo Document of 2005) led to the revised international treaty's abandoning of specific references to CBRN in favor of an all-risks scope (which could still be interpreted to include CBRN) (Weir 2014).

19. The first document was a May 2005 guidance document outlining actions for the WHO and recommendations for national authorities to implement flu preparedness; it updated the preliminary pandemic flu guidance plan of 1999, and included planning and coordination, situation monitoring and assessment, prevention and containment, health system response, and communications (WHO 2005e).

20. More specifically, clinical specimens such as throat and nasal swabs, endotracheal aspirates, and lung biopsies can be tested to identify the viruses' RNA structures and thereby produce updated vaccines.

21. Communication studies scholar Nina Song (2007) has noted the way U.S. media frequently articulated China's handling of H5N1 by referencing its past handling of SARS.

22. With increased Chinese immigration in the late 1800s and the threat Chinese settlement posed to white labor, white society increasingly viewed Chinese people as an inassimilable, immoral, and disease-ridden "yellow plague" (E. Lee 2007; Shah 2001).

23. China still faced Western powers' continued neocolonial practices and the legacy of semicolonialism: eighteen foreign powers vied over China from the mid-1800s to the mid-1900s, creating a partial, multiple, and layered colonial experience. The colonial formations that resulted were geographically limited to coastal cities only; they were also limited in being unable to assume formal sovereignty over China (Shih 2001).

24. See Chen et al. (2007) for more on investigations into the role of Chinese public health (particularly traditional Chinese medicine) in lowering fatality rates.

25. Some scholars argued that Chinese governmental secretiveness (in Beijing or in Guangdong or in both) led to the spread of SARS (Huang 2004; White 2003) or that the poor health care systems in rural parts of China were culpable for spread (Davis and Siu 2006; Kaufman 2006). Others mostly praised the Chinese public health response, particularly in Guangdong, and only criticized the vast

decentralized national system as disorganized and unmanageable (Schnur 2006); and still others emphasized the ability of the Chinese public health system to quickly gain control of a new epidemic—with mass mobilization and successful Chinese public health campaigns—only three months after the disease's emergence (Kaufman 2006; Lee and Warner 2008).

26. As Edward Said (1978) famously noted, knowledge production about the "Orient" constructs the countries and peoples under its rubric in ways that help maintain Western colonial power over the "Orient."

27. U.S. portrayals of China during H5N1 continued themes that emerged during SARS: criticism of East Asian eating and farming practices, general population density, and inadequate public health measures were viewed as responsible for flu etiology and spread (Respiratory Diseases Committee of the American Association of Avian Pathologists n.d.; Zamiska and Champion 2006). U.S. pundits seldom emphasized the lack of information on factors of pandemic influenza's emergence and spread, or that many scientists attribute the cause of pandemics to industrialization and factory farming (Greger 2006; Schrijver and Koch 2005).

28. This is both because such border-based methods are nearly impossible to implement in an increasingly interconnected world and because people infected with flu viruses are contagious for days before their symptoms show. Even border screening for a disease like SARS, where symptoms show days before peak infectivity, had proven ineffective (St. John et al. 2005).

29. The reality was that Asian countries have engaged in a high degree of practices of international health cooperation in the twenty-first century (particularly China, India, and Japan), including bilateral relations, regional activities, and participation in multilateral organizations (Fidler 2010).

30. The 2005 IHR lacked an enforcement mechanism, and there was some room for interpretation as to the conditions under which the 2005 IHR could be applied (Fidler and Gostin 2006; Lakoff 2010; WHO 2005c).

31. Davies examined government reports to the WHO as well as public information on outbreaks recorded on ProMED-Mail, since the latter proved highly accurate in later verification of the disease events. She

further revealed that the international community applied scrutiny unevenly—criticizing China, India, and Thailand as failing to report promptly, but not Vietnam, which she showed had significant reporting gaps in communication with the WHO (Davies 2012).

32. In 2011 it was renamed the Global Influenza Surveillance and Response System (GISRS).

33. Smallman (2013) focuses on the 2009 H1N1 pandemic as the denouement of this inequity. Unlike H5N1, H1N1 had a high enough incidence to activate vaccine distribution—death tolls numbered over 18,036 in over 214 countries by conservative estimates (WHO 2010); Australia, Canada, and the Netherlands were able to acquire the vaccine, but heavily impacted nations such as Mexico had to wait for wealthy nations to share excess vaccines with them. Almost all the first billion doses went to twelve wealthy nations that had made advance orders; moreover, the 120 million doses Sanofi Pasteur and GlaxoSmithKline pledged to the WHO for distribution to poor countries could only be fulfilled months after the pandemic had waned.

34. As mentioned in chapter 2, vaccines are not the most lucrative of pharma products because they are single-use.

35. These included a press conference on February 7, 2007, announcing the pact Indonesia signed with Baxter to develop the H5N1 vaccine; a March 2007 speech at the high-level WHO meeting held in Jakarta titled "Responsible Practices for Sharing Avian Influenza Viruses and Resulting Benefits"; the WHO Intergovernmental Meeting on Pandemic Influenza Preparedness in November 2007; Supari's early 2008 book *It's Time for the World to Change: Divine Hand behind Avian Influenza*; and a multiple-authored article from the Indonesian Ministry of Health published in the *Annals of the Academy of Medicine* in June 2008.

36. The guideline, instituted in March 2005, in *Guidance for the Timely Sharing of Influenza Viruses/Specimens with Potential to Cause Human Influenza Pandemics*, states, "There shall be no further distribution of viruses/specimens outside the network WHO Reference Laboratories without permission from the originating country/laboratory" (WHO 2005a). Tellingly, it was later overruled. See Vezzani 2010 for details.

37. Harriet Washington's chapter "Biocolonialism," in her book *Deadly Monopolies: The Shocking Corporate Takeover of Life Itself—and the Consequences for Your Health and Our Medical Future* (2012), comprehensively details "the appropriation of the biological riches (plants, animals, bodies, tissues, genes, etc.) of the poor, the marginal, the weak, the subjugated, and the genetically distinct for the Western medical marketplace." Vandana Shiva (1997) has long critiqued the Global North's commodification of environmental, food, and health resources in the Global South and its harmful effects on local ecologies and local labor forces—often women—who work closely with the natural environment. The Indigenous Peoples Council on Biocolonialism, founded in 1999 and based in Nevada, does prominent work on the ground against bio-imperialism; its mission is to "assist indigenous peoples in the protection of their genetic resources, indigenous knowledge, cultural and human rights from the negative effects of biotechnology" (Indigenous Peoples Council on Biocolonialism n.d.).

38. Laurelyn Whitt, who focuses on plant resources specifically, uses the term to highlight how this extractive process differs from other forms of biocolonialism such as the introduction of monocultures and resultant undermining of plant genetic diversity. She also notes, as do I, that "biocolonialism" is a more apt term than the frequently used term "biopiracy," because the latter limits the issue to an *act* of misappropriation and abuse, whereas the former signals a larger *system* of oppressive relations between the West/Global North and indigenous or marginalized cultures.

39. "Biocolonialism" has seldom been applied to describe the way flu-afflicted regions have had to hand over their flu data and samples to the international health system. A search in 2017 in the ProQuest databases yielded only two articles that utilized "biocolonialism" to refer to the acquisition of germ data/samples during the H5N1 flu sharing controversy: one a short *BMC Medical Ethics* journal article (Emerson, Singer, and Upshur 2011); the other a review article in the *African Journal of Political Science and International Relations* (Mukhopadhyay 2013).

40. The 1992 Convention on Biological Diversity is a multilateral treaty that seeks the conservation of biological diversity, the sustainable use

of its components, and the fair and equitable sharing of benefits arising from genetic resources.

41. While the WHO had made some concessions early on, such as offering Indonesia laboratory improvements and free vaccine in February 2007, Indonesia had chosen to fight for a revised WHO research system and greater support for improving the production capacity of vaccines in Indonesia and other impacted nations (Elbe 2010; Lin 2015).

Epilogue

1. The Obama administration overhauled lab biosecurity regulations to make them more efficient, creating a tiered system of priority (encoded in the WMD Prevention and Preparedness Act of 2009).

2. Al Qaeda fighters claimed that the Syrian government had attacked them with chemical weapons. The veracity of these claims is contested (Cook 2019; OPCW 2019).

3. In April 2014 the University of Massachusetts–Amherst hosted a conference, "Science for the People: The 1970s and Today," which sparked interest in the organization's revitalization (Science for the People n.d.a).

4. These working groups include Climate Change; Nuclear Disarmament; Militarism; Labor; Biology and Society; STEM Intellectuals under Attack; Reproductive Justice; Science Education and Social Responsibility; Science for Puerto Rico Group; and Technology (Science for the People n.d.e).

5. By August 2010, the U.S. Defense Department had shifted more than $1 billion out of its nuclear, biological, and chemical defense programs to underwrite a new White House priority to combat disease pandemics with vaccine development and production (Nuclear Threat Initiative 2010).

6. H1N1 led to estimated hundreds of thousands of deaths worldwide (and tens of thousands in the United States) by the time of its decline in mid-2010 (CDC 2012; Shrestha et al. 2011).

7. In 2012, a Dutch university that had received MERS virus samples from Saudi Arabia delayed sending the virus to other laboratories for a few weeks while it prepared its own patent application, providing the

virus only after it had devised a material transfer agreement reserving intellectual property rights for the university; and in 2019, scientists from Sierra Leone, Guinea, and Liberia were unable to access Ebola samples that the WHO had extracted during the 2014–2016 crisis—the samples were shared with laboratories globally, but as of this writing African scientists cannot access them (Third World Network 2019).

8. Formed in 1984, the Third World Network seeks to address the needs and rights of peoples in the South, achieve a fair distribution of world resources, and implement forms of development that are ecologically sustainable and fulfill human needs. In November 2007, the Network put out a statement outlining its support of the Global South's proposed changes to the global flu system (Third World Network 2007a).

9. Nations had been wrangling over the principle of access and benefit sharing of biological resources since its origins in the 1992 Convention on Biological Diversity (CBD), but even more so with the implementation of the 2011 Pandemic Influenza Preparedness Framework and the addition to the CBD of the 2010 Nagoya Protocol on Access to Genetic Resources and the Fair and Equitable Sharing of Benefits Arising from Their Utilization, which went into effect on October 12, 2014.

10. United American Nurses was a union affiliated with the American Federation of Labor and Congress of Industrial Organizations (AFL-CIO).

11. The Trump administration's proposed 2018 budget slated cuts for funds to the NIH, disease research, and early disease warning systems such as BioWatch. The administration, moreover, outlined a long-term vision to severely curtail health spending over ten years (Machi 2017; Newkirk 2017; Scott 2018).

References

The 2009 Influenza Pandemic: An Overview. 2009. R40554. Congressional Research Service. https://www.everycrsreport.com/reports/R40554 .html#ifn20.

AAAS, Human Rights Action Network. 2005. "Iraqi Scientist Being Held without Charges or Trial." August 11. Accessed November 10, 2007. https://www.aaas.org/programs/scientific-responsibility-human-rights -law.

Abraham, Thomas. 2011. "The Chronicle of a Disease Foretold: Pandemic H1N1 and the Construction of a Global Health Security Threat." *Political Studies* 59 (4): 797–812. https://doi.org/10.1111/j.1467-9248.2011 .00925.x.

Adas, Michael. 1989. *Machines as the Measure of Men: Science, Technology, and Ideologies of Western Dominance.* Ithaca, NY: Cornell University Press.

Ahuja, Neel. 2016. "Staging Smallpox: Reanimating Variola in the Iraq War." In *Bioinsecurities: Disease Interventions, Empire, and the Government of Species*, 133–168. Durham, NC: Duke University Press. www .dukeupress.edu/bioinsecurities.

Alberts, Bruce, and Robert M. May. 2002. "Scientist Support for Biological Weapons Controls." *Science* 298 (5596): 1135. https://doi.org/10.1126 /science.298.5596.1135.

Aldhous, Peter, and Michael Reilly. 2006. "Bioterror Special: Friend or Foe?" *New Scientist*, October 11.

Aldis, William S., and Triono Soendoro. 2014. "Indonesia, Power Asymmetry, and Pandemic Risk." In *Routledge Handbook of Global*

Health Security, edited by Simon Rushton and Jeremy R. Youde, 318–327. London: Routledge.

Alsultany, Evelyn. 2008. "The Prime-Time Plight of the Arab Muslim American after 9/11: Configurations of Race and Nation in TV Dramas." In *Race and Arab Americans before and after 9/11: From Invisible Citizens to Visible Subjects*, edited by Amaney Jamal and Nadine Naber, 204–228. Syracuse, NY: Syracuse University Press.

Althusser, Louis. 1971. "Ideology and State Apparatuses (Notes towards an Investigation)." *Lenin and Philosophy and Other Essays*, translated by B. Brewster, 171–174. New York: Monthly Review Press.

Amar, Paul. 2011. "Middle East Masculinity Studies: Discourses of 'Men in Crisis,' Industries of Gender in Revolution." *Journal of Middle East Women's Studies* 7 (3): 36–70.

Amireh, Amal. 2011. "Palestinian Women's Disappearing Act: The Suicide Bomber through Western Feminist Eyes." In *Arab and Arab American Feminisms: Gender, Violence, and Belonging*, edited by Rabab Abdulhadi, Evelyn Alsultany, and Nadine Naber, 29–45. Syracuse, NY: Syracuse University Press.

Anderson, Ben. 2010. "Preemption, Precaution, Preparedness: Anticipatory Action and Future Geographies." *Progress in Human Geography* 34 (6): 777–798. https://doi.org/10.1177/0309132510362600.

———. 2012. "Affect and Biopower: Towards a Politics of Life." *Transactions of the Institute of British Geographers* 37 (1): 28–43. https://doi.org/10.1111/j.1475-5661.2011.00441.x.

Anderson, Brian K. 1999. "A Profile of WMD Proliferants: Are There Commonalities?" Maxwell Air Force Base, Alabama. https://fas.org/irp/threat/99-003.pdf.

Anderson, Warwick. 2006. *Colonial Pathologies: American Tropical Medicine, Race, and Hygiene in the Philippines*. Durham, NC: Duke University Press.

Asad, Talal. 2007. *On Suicide Bombing*. New York: Columbia University Press.

Atlas, Ronald M., and Malcolm Dando. 2006. "The Dual-Use Dilemma for the Life Sciences: Perspectives, Conundrums, and Global Solutions." *Biosecurity and Bioterrorism: Biodefense Strategy, Practice, and Science* 4 (3): 276–286. https://doi.org/10.1089/bsp.2006.4.276.

Barnaby, Wendy. 2000. *Plague Makers: The Secret World of Biological Warfare*. London: Continuum.

Beck, Lindsay. 2006. "China Shares Bird Flu Samples, Denies New Strain Report." *Reuters*, November 10.

Bernstein, B. J. 1987. "The Birth of the U.S. Biological-Warfare Program." *Scientific American* 256 (6): 116–121.

Bhattacharya, Shaoni. 2003. "Bioterrorist Fears Prompt Journal Paper Censorship." *New Scientist*, February 17. https://www.newscientist .com/article/dn3394-bioterrorist-fears-prompt-journal-paper -censorship/.

Bhattacharyya, Gargi. 2008. *Dangerous Brown Men: Exploiting Sex, Violence and Feminism in the "War on the Terror."* London: Zed Books.

Bhuiyan, A.J.M. 2008. "Peripheral View: Conceptualizing the Information Society as a Postcolonial Subject." *International Communication Gazette* 70 (2): 99–116.

Biological Weapons Convention. 1972. http://www.opbw.org/convention/conv .html.

Black, Lee. 2007. "Informed Consent in the Military: The Anthrax Vaccination Case." *AMA Journal of Ethics* 9 (10): 698–702. https://doi .org/10.1001/virtualmentor.2007.9.10.hlaw1-0710.

Boyd, J. Wesley. 2001. "The War over Here: Anthrax Spreads Its Terror." *New York Times*, October 18, 2001, Opinion sec.

Bresalier, Michael. 2012. "Sharing Viruses and Vaccines: Economies of Exchange in Global Influenza Control since 1947." Institut d'Anatomie pathologique, Université de Strasbourg. https://www.academia.edu /3644403/Sharing_viruses_and_vaccines_Economies_of_exchange_in _global_influenza_control_since_1947.

Briggs, Charles L., and Mark Nichter. 2009. "Biocommunicability and the Biopolitics of Pandemic Threats." *Medical Anthropology* 28 (3): 189–198. https://doi.org/10.1080/01459740903070410.

Broad, William J. 2002. "Bush Signals He Thinks Possibility of Smallpox Attack Is Rising." *New York Times*, December 14, 2002, National edition, Politics sec. https://www.nytimes.com/2002/12/14/politics /bush-signals-he-thinks-possibility-of-smallpox-attack-is-rising.html.

———. 2003. "Natural Causes Emerge as Key to Mystery Illness." *New York Times*, April 6, late edition, sec. A.

Broad, William J., and Melody Petersen. 2001. "A Nation Challenged: The Biological Threat; Nation's Civil Defense Could Prove to Be Inadequate against a Germ or Toxic Attack." *New York Times*, September 23, 2001. https://www.nytimes.com/2001/09/23/us/nation-challenged-biological -threat-nation-s-civil-defense-could-prove-be.html.

Brookes, Timothy J., and Omar A. Khan. 2005. *Behind the Mask: How the World Survived SARS, the First Epidemic of the 21st Century*. Washington, DC: American Public Health Assn.

Brown, Theodore M., Marcos Cueto, and Elizabeth Fee. 2006. "The World Health Organization and the Transition from 'International' to 'Global' Public Health." *American Journal of Public Health* 96 (1): 62–72. https://doi.org/10.2105/AJPH.2004.050831.

Brunner, Claudia. 2007. "Occidentalism Meets the Female Suicide Bomber: A Critical Reflection on Recent Terrorism Debates; A Review Essay on JSTOR." *Signs: Journal of Women in Culture and Society* 32 (4): 957–971.

Budget of the U.S. Government. 2003. Government Printing Office. https://www.govinfo.gov/content/pkg/BUDGET-2003-BUD/pdf /BUDGET-2003-BUD.pdf.

Bureau of Public Affairs. 2007. "United States International Engagement on Avian and Pandemic Influenza." Department of State, Office of Electronic Information. https://2001-2009.state.gov/r/pa/scp/88620.htm.

BurrellesLuce. 2007. "Top Newspapers, Blogs & Consumer Magazines." Accessed March 28, 2008. https://burrelles.com/wp-content/uploads /2017/12/2007_Top_100List.pdf.

Bush, George W. 2001a. "Address to a Joint Session of Congress and the American People." Office of the Press Secretary, Washington, DC, September 20. https://georgewbush-whitehouse.archives.gov/news /releases/2001/09/20010920-8.html.

———. 2001b. "'Islam Is Peace' Says President." Islamic Center of Washington, DC, September 17. https://georgewbush-whitehouse.archives .gov/news/releases/2001/09/20010917-11.html.

———. 2001c. "President Discusses War on Terrorism." Presented at the Address to the Nation, Atlanta, Georgia, November 8. https:// georgewbush-whitehouse.archives.gov/news/releases/2001/11/20011108 -13.html.

————. 2001d. "Remarks to the United Nations General Assembly in New York City, November 10, 2001." 46. *Weekly Compilation of Presidential Documents* 37.

————. 2002a. "President Bush Outlines Iraqi Threat." Cincinnati Museum Center, Cincinnati Union Terminal, October 7. https://georgewbush -whitehouse.archives.gov/news/releases/2002/10/20021007-8.html.

————. 2002b. "President Delivers Remarks on Smallpox." Press release, Dwight D. Eisenhower Executive Office Building, December 13. https://georgewbush-whitehouse.archives.gov/news/releases/2002/12 /20021213-7.html.

————. 2002c. "President Signs Public Health Security and Bioterrorism Bill." Office of the Press Secretary, June 12. https://georgewbush -whitehouse.archives.gov/news/releases/2002/06/20020612-1.html.

————. 2004a. "Fact Sheet: President Bush Signs Biodefense for the 21st Century." Office of Press Secretary. https://www.hsdl.org/ ?abstract&did=476622.

————. 2004b. "Homeland Security Presidential Directive: Biodefense for the 21st Century." The White House: Office of the Press Secretary. https://www.hsdl.org/?view&did=784400.

————. 2005. "President Outlines Pandemic Influenza Preparations and Response." Bethesda, MD, November. https://georgewbush -whitehouse.archives.gov/news/releases/2005/11/20051101-1.html.

Cainkar, Louise. 2008. "Thinking outside the Box." In *Race and Arab Americans before and after 9/11: From Invisible Citizens to Visible Subjects,* edited by Amaney Jamal and Nadine Naber, 46–80. Syracuse, NY: Syracuse University Press.

Calain, Philippe. 2007a. "Exploring the International Arena of Global Public Health Surveillance." *Health Policy and Planning* 22 (1): 2–12. https://doi.org/10.1093/heapol/czl034.

————. 2007b. "From the Field Side of the Binoculars: A Different View on Global Public Health Surveillance." *Health Policy and Planning* 22 (1): 13–20. https://doi.org/10.1093/heapol/czl035.

California Nurses Association. 2003. "CNA Adds Voice to Opposition to Smallpox Vaccination Plan." *California Nurse* 99 (1/2): 5.

Cameron, Lynne. 2009. "Responding to the Risk of Terrorism: The Contribution of Metaphor." *DELTA: Documentação de Estudos Em*

Lingüística Teórica e Aplicada 26 (SPE): 587–614. https://doi.org/10.1590
/S0102-44502010000300010.

Caplan, Bryan. 2006. "Terrorism: The Relevance of the Rational Choice
Model." *Public Choice* 128 (1–2): 91–107. https://doi.org/10.1007/s11127
-006-9046-8.

Carus, Seth W. 2001. "Bioterrorism and Biocrimes: The Illicit Use of
Biological Agents in the 20th Century." Washington, DC: Center for
Counterproliferation Research, National Defense University.

Casey, C. G., J. Iskander, and M. H. Roper. 2005. "Adverse Events
Associated with Smallpox Vaccination: Results from the US Department of Health and Human Services Smallpox Vaccine Safety
Monitoring and Response System, January–October 2003." *JAMA* 294
(21): 2734–2743.

CDC. *See* Centers for Disease Control and Prevention

Cecire, Ruth. 2009. "Bioweapons: Postmodern Ruminations on a
Premodern Modality." *Feminist Studies* 35 (1): 41–65.

Center for Biosecurity of UPMC. 2006. "Comments from the Center for
Biosecurity of UPMC on the National Strategy for Pandemic Influenza:
Implementation Plan." *Biosecurity and Bioterrorism: Biodefense Strategy,
Practice, and Science* 4 (3): 320–324. https://doi.org/10.1089/bsp.2006.4.320.

Center for Counterproliferation Research. 2002. "Anthrax in America: A
Chronology and Analysis of the Fall 2001 Attacks." Washington, DC:
National Defense University.

———. 2003. "Toward a National Biodefense Strategy: Challenges and
Opportunities; A Report of the Center for Counterproliferation
Research." National Defense University. https://www.hsdl.org/
?abstract&did=3596.

Center for Public Integrity. 2008. "Key False Statements." Center for
Public Integrity. January 23. https://www.publicintegrity.org/2008/01
/23/5644/key-false-statements.

Centers for Disease Control and Prevention. n.d.a. "Highly Pathogenic Asian
Avian Influenza A (H5N1) in People." Avian Influenza (Flu). Accessed
June 11, 2015. https://www.cdc.gov/flu/avianflu/h5n1-people.htm.

———. n.d.b. "Highly Pathogenic Asian Avian Influenza A (H5N1)
Virus." Accessed April 27, 2015. https://www.cdc.gov/flu/avianflu/h5n1
-virus.htm.

———. n.d.c. "Novel H1N1 Vaccination Recommendations." H1N1 Flu. Accessed October 15, 2009. https://www.cdc.gov/h1n1flu/vaccination /acip.htm.

———. n.d.d. "Past Avian Influenza Outbreaks." Avian Influenza (Flu). Accessed December 1, 2006. https://www.cdc.gov/flu/avianflu/past -outbreaks.htm.

———. 2002. "Record of the Meeting of the Advisory Committee on Immunization Practices." June 19–20.

———. 2003a. "CDC Lab Analysis Suggests New Coronavirus May Cause SARS." March 24. https://www.cdc.gov/media/pressrel/r030324 .htm.

———. 2003b. "CDC Telebriefing Transcript—Safer, Healthier Summer." CDC Media Relations. June 26. https://www.cdc.gov/media /transcripts/t030626.htm.

———. 2003c. "Recommendations for Using Smallpox Vaccine in a Pre-event Vaccination Program." 52(RR07). MMWR. April 4. https://www.cdc.gov/mmwr/preview/mmwrhtml/rr5207a1.htm.

———. 2003d. "Update: Severe Acute Respiratory Syndrome—United States, 2003." 52(RR15). MMWR. April 18. https://www.cdc.gov /mmwr/preview/mmwrhtml/mm5215a2.htm.

———. 2012. "First Global Estimates of 2009 H1N1 Pandemic Mortality Released by CDC-Led Collaboration." June 25. https://www.cdc.gov /flu/spotlights/pandemic-global-estimates.htm.

Central Intelligence Agency. 2007. *Iraq's Weapons of Mass Destruction Programs.* https://www.cia.gov/library/reports/general-reports-1/iraq _wmd/Iraq_Oct_2002.htm#06.

Chan-Malik, Sylvia. 2011. "Chadors, Feminists, Terror: The Racial Politics of U.S. Media Representations of the 1979 Iranian Women's Move-ment." *Annals of the American Academy of Political and Social Science* 637 (1): 112–140. https://doi.org/10.1177/0002716211409011.

Chase, Marilyn. 2001. "Demand Grows for Vaccinations against Small-pox." *Wall Street Journal*, November 6, 2001, Eastern Edition.

Chen, Yan, Jeff J. Guo, Daniel P. Healy, and Siyan Zhan. 2007. "Effect of Integrated Traditional Chinese Medicine and Western Medicine on the Treatment of Severe Acute Respiratory Syndrome: A Meta-Analysis." *Pharmacy Practice* 5 (1): 1–9.

China's Response to Avian Flu: Steps Taken, Challenges Remaining, and Transparency: Hearing before the Congressional-Executive Commission on China. 2006. 109th Cong., 1st sess. (February 24). https://www.govinfo .gov/content/pkg/CHRG-109hhrg26672/html/CHRG-109hhrg26672 .htm.

Chow, Rey. 2002. "From Biopower to Ethnic Difference." In *The Protestant Ethnic and the Spirit of Capitalism*, 1–18. New York: Columbia University Press.

Christopher, George W., Theodore J. Cieslak, Julie A. Pavin, and Edward M. Eitzen Jr. 1997. "Biological Warfare: A Historical Perspective." *Journal of the American Medical Association* 278: 412–417.

Chyba, Christopher F. 2002. "Toward Biological Security." *Foreign Affairs*, May 1. https://www.foreignaffairs.com/articles/2002-05-01/toward -biological-security.

Clarke, Robin. 1968. *The Silent Weapons: The Realities of Chemical and Biological Warfare.* 1st ed. New York: David McKay Co.

Clinton, Bill. 1998. "Text of Clinton Statement on Iraq." February 17. http://www.cnn.com/ALLPOLITICS/1998/02/17/transcripts/clinton .iraq/.

———. 1999. "Keeping America Secure for the 21st Century." Presented at the National Academy of Sciences, Washington, DC, January 22. https://clintonwhitehouse5.archives.gov/WH/new/html/19990122-7214 .html.

"Clinton Administration Declares AIDS a New Threat to National Security and Global Stability." 2000. CNN. http://edition.cnn.com /TRANSCRIPTS/0004/30/sun.02.html.

Cohen, Hillel W., Robert M. Gould, and Victor W. Sidel. 2004. "The Pitfalls of Bioterrorism Preparedness: The Anthrax and Smallpox Experiences." *American Journal of Public Health* 94 (10): 1667–1671.

Cohen, Jon. 2001. "Smallpox Vaccinations: How Much Protection Remains?" *Science* 294 (5544): 985.

Cohen, Jon, and Martin Enserink. 2002. "Public Health: Rough-and-Tumble behind Bush's Smallpox Policy." *Science* 298 (5602): 2312–2316.

Cohn, Carol. 1987. "Sex and Death in the Rational World of Defense Intellectuals." *Signs: Journal of Women in Culture and Society* 12 (4): 687–718. https://doi.org/10.1086/494362.

Cole, Alyson Manda. 2007. *The Cult of True Victimhood: From the War on Welfare to the War on Terror.* Palo Alto, CA: Stanford University Press.

Cole, Leonard A. 1997. *The Eleventh Plague: The Politics of Biological Chemical Warfare.* 1st ed. New York: W. H. Freeman.

———. 2003. *The Anthrax Letters: A Medical Detective Story.* Washington, DC: Joseph Henry Press.

———. 2016. "Open-Air Biowarfare Testing and the Evolution of Values." *Health Security* 14 (5): 315–322. https://doi.org/10.1089/hs .2016.0040.

Cook, Jonathan. 2019. "The Western Media Is Key to Syria Deceptions." *Common Dreams*, May 27. https://www.commondreams.org/views/2019 /05/27/western-media-key-syria-deceptions.

Cooper, Melinda E. 2008. *Life as Surplus: Biotechnology and Capitalism in the Neoliberal Era.* Seattle: University of Washington Press.

Corrigall-Brown, Catherine. 2011. "The Power of Pictures: Images of Politics and Protest." *American Behavioral Scientist* 56 (2): 131–134.

Crenshaw, Martha. 1990. "The Logic of Terrorism: Terrorist Behavior as a Product of Strategic Choice." In *Origins of Terrorism: Psychologies, Ideologies, Theologies, States of Mind*, 1st ed., 7–24. Baltimore: Johns Hopkins University Press.

Crook, John R., ed. 2006. "United States Initiates International Partnership on Avian and Pandemic Influenza." *American Journal of International Law* 100 (1): 226–227. https://doi.org/10.2307/3518848.

Curtis, Edward E. 2013. "The Black Muslim Scare of the Twentieth Century." In *Islamophobia in America: The Anatomy of Intolerance*, edited by C. Ernst, 75–106. New York: Palgrave Macmillan.

Da Costa, Beatriz. 2010. "Amateur Science, a Threat after All?" http:// nideffer.net/shaniweb/files/threat.pdf.

Dando, Malcolm. 2006. *Bioterror and Biowarfare: A Beginner's Guide.* Oxford: Oneworld Publications.

D'Antonio, Patricia. 2010. "Histories of Nursing: The Power and the Possibilities." *Nursing Outlook* 58 (4): 207–213. https://doi.org/10.1016/j .outlook.2010.04.005.

D'Arcangelis, Gwen. 2008. "Chinese Chickens, Ducks, Pigs, and Humans, and the Technoscientific Discourses of Global U.S. Empire—MIT Press Scholarship." July 3. https://www.academia.edu

/12234023/Chinese_chickens_ducks_pigs_and_humans_and_the
_Technoscientific_Discourses_of_Global_U.S._Empire.

———. 2009. *The Bio Scare: Anthrax, Smallpox, SARS, Flu and Post-9/11 U.S. Empire.* PhD diss. University of California. https://books.google .com/books?id=qUPkSAAACAAJ.

———. 2015. "Defending White Scientific Masculinity: The FBI, the Media and Profiling Tactics during the Post-9/11 Anthrax Investigation." *International Feminist Journal of Politics* 18 (1): 1–20. https://doi .org/10.1080/14616742.2015.1051330.

———. 2017. "Reframing the 'Securitization of Public Health': A Critical Race Perspective on Post-9/11 Bioterrorism Preparedness in the US." *Critical Public Health* 27 (2): 275–284. https://doi.org/10.1080/09581596 .2016.1209299.

———. 2019. "Confronting Public Health Imperialism: A Transnational Feminist Analysis of Critical Nurse Responses to the National Smallpox Vaccination Program of 2002." *Frontiers: A Journal of Women Studies* 40 (1): 95–121. https://muse.jhu.edu/article/719765.

Davies, Sara E. 2008. "Securitizing Infectious Disease." *International Affairs* 84 (2): 295–313. https://doi.org/10.1111/j.1468-2346.2008.00704.x.

———. 2012. "The International Politics of Disease Reporting: Towards Post-Westphalianism?" *International Politics* 49 (5): 591–613. https://doi .org/10.1057/ip.2012.19.

Davis, Deborah, and Helen F. Siu, eds. 2006. *SARS: Reception and Interpretation in Three Chinese Cities.* London: Routledge.

Department of Defense. n.d. "Casualty Status." Accessed August 7, 2019. https://www.defense.gov/Newsroom/Casualty-Status/.

———. 2006. *Implementation Plan for Pandemic Influenza.* Washington, DC: Homeland Defense. https://www.hsdl.org/?view&did=473250.

Department of Health and Human Services. 2005. *Pandemic Influenza Plan.* https://www.cdc.gov/flu/pdf/professionals /hhspandemicinfluenzaplan.pdf.

Department of Justice. 2010. "Amerithrax Investigative Summary." February 19. https://www.justice.gov/archive/amerithrax/docs/amx -investigative-summary.pdf.

Department of State. 2002. "Overview of State-Sponsored Terrorism." Patterns of Global Terrorism. Office of the Coordinator for

Counterterrorism, May 21. https://2009-2017.state.gov/j/ct/rls/crt/2001
/html/10249.htm.

———. 2005. "Safeguarding America against Pandemic Influenza." Office
of Electronic Information. https://2001-2009.state.gov/g/oes/rls/fs
/55881.htm.

DiAngelo, Robin. 2018. *White Fragility: Why It's So Hard for White People to
Talk about Racism*. Reprint. Boston: Beacon Press.

Dias, M. Beatrice, Leonardo Reyes-Gonzalez, Francisco M. Veloso, and
Elizabeth A. Casman. 2010. "Effects of the USA PATRIOT Act and
the 2002 Bioterrorism Preparedness Act on Select Agent Research in
the United States." *PNAS* 107 (21): 9556–9561.

Dorsey, M. K. 2004. "Managing, Manipulating and Maneuvering Biology
in the Early 21st Century: Reflections on Discursive Practice, Empiri-
cal Events and Power in Pursuit of Bio-Commerce." *SSRN Electronic
Journal*, December. https://doi.org/10.2139/ssrn.643401.

Dowd, Maureen. 2001. "From Botox to Botulism." *New York Times*,
September 26, 2001, late edition, sec. A.

Duffy, John. 2002. "Smallpox and the Indians in the American Colonies."
In *Bioterrorism: The History of a Crisis in American Society*, edited by
David McBride. Vol. 1. New York: Routledge.

Eagan, Catherine M. 2003. "White, if 'Not Quite': Irish Whiteness in the
Nineteenth-Century Irish-American Novel." In *New Directions in
Irish-American History*, edited by Kevin Kenny, 140–155. Madison:
University of Wisconsin Press.

"Editors' Statement on Considerations of Biodefence and Biosecurity."
2003. *Nature Medicine* 9 (3): 240. https://doi.org/10.1038/nm0303-240.

Elbe, Stefan. 2010. "Haggling over Viruses: The Downside Risks of
Securitizing Infectious Disease." *Health Policy and Planning* 25 (6):
476–485. https://doi.org/10.1093/heapol/czq050.

Emerson, Claudia, Peter A. Singer, and Ross E. G. Upshur. 2011. "Access
and Use of Human Tissues from the Developing World: Ethical
Challenges and a Way Forward Using a Tissue Trust." *BMC Medical
Ethics* 12 (2): 1–5.

Enloe, Cynthia. 2014. *Bananas, Beaches and Bases: Making Feminist Sense
of International Politics*. 2nd ed. Berkeley: University of California
Press.

Enserink, Martin. 2002. "How Devastating Would a Smallpox Attack Really Be?" *Science* 296 (5573): 1592–1595. https://doi.org/10.1126/science.296.5573.1592.

———. 2005. "WHA Gives Yellow Light for Variola Studies." *Science* 308 (5726): 1235. https://doi.org/10.1126/science.308.5726.1235.

Escobar, Arturo. 1995. *Encountering Development: The Making and Unmaking of the Third World*. Princeton, NJ: Princeton University Press.

Everett, Worth W., Susan E. Coffin, Theoklis Zaoutis, Scott D. Halpern, and Brian L. Strom. 2003. "Smallpox Vaccination: A National Survey of Emergency Health Care Providers." *Academic Emergency Medicine: Official Journal of the Society for Academic Emergency Medicine* 10 (6): 606–611.

Falk, Richard. 1990. "Inhibiting Reliance on Biological Weaponry: The Role and Relevance of International Law." In *Preventing a Biological Arms Race*, edited by Susan Wright, 241–266. Boston: MIT Press.

Fedson, David S. 2003. "Pandemic Influenza and the Global Vaccine Supply." *Clinical Infectious Diseases* 36 (12): 1552–1561.

Fidler, David P. 2002. "Bioterrorism, Public Health, and International Law." *Chicago Journal of International Law* 3 (1): 7–26.

———. 2005. *SARS, Governance and the Globalization of Disease*. Palgrave Macmillan.

———. 2007. "Indonesia's Decision to Withhold Influenza Virus Samples from the World Health Organization: Implications for International Law." *American Society for International Law* 11 (4). https://www.asil.org/insights/volume/11/issue/4/indonesias-decision-withhold-influenza-virus-samples-world-health.

———. 2008. "Influenza Virus Samples, International Law, and Global Health Diplomacy." *Emerging Infectious Diseases* 14 (1): 88–94. https://www.ncbi.nlm.nih.gov/pubmed/18258086.

———. 2010. "Asia's Participation in Global Health Diplomacy and Global Health Governance." *Asian Journal of WTO & International Health Law and Policy* 5 (2): 269–300. https://www.researchgate.net/publication/228122912_Asia's_Participation_in_Global_Health_Diplomacy_and_Global_Health_Governance.

Fidler, David P., and Lawrence O. Gostin. 2006. "The New International Health Regulations: An Historic Development for International Law and Public Health." *Journal of Law, Medicine and Ethics* 34 (1): 85–94. https://doi.org/10.1111/j.1748-720X.2006.00011.x.

———. 2007. *Biosecurity in the Global Age: Biological Weapons, Public Health, and the Rule of Law.* Stanford, CA: Stanford Law and Politics.

Findlay, Trevor. 2006. "Verification and the BWC: Last Gasp or Signs of Life?" *Arms Control Today*, September. https://www.armscontrol.org/act/2006_09/BWCVerification.

Foucault, Michel. 1977. *Discipline and Punish.* New York: Vintage.

———. 1978. *The History of Sexuality: An Introduction.* New York: Vintage.

———. 1986. *The Care of the Self.* New York: Vintage.

Franklin, Nicole. 2009. "Sovereignty and International Politics in the Negotiation of the Avian Influenza Material Transfer Agreement." *Journal of Law and Medicine* 17 (3): 355–372.

Fraser, Nancy. 2003. "From Discipline to Flexibilization? Rereading Foucault in the Shadow of Globalization." *Constellations* 10 (2): 160–171. https://doi.org/10.1111/1467-8675.00321.

Garrett, Laurie. 1995. *The Coming Plague: Newly Emerging Diseases in a World out of Balance.* London: Penguin.

Gaudioso, Jennifer, and Reynolds M. Salerno. 2004. "Biosecurity and Research: Minimizing Adverse Impacts." *Science* 304 (5671): 687. https://doi.org/10.1126/science.1096911.

Gellene, Denise. 2003. "Homeland Security Becomes an Opportunity for Biotech Firms." *Los Angeles Times*, March 3, 2003, California edition.

Gellman, Barton. 2000. "AIDS Is Declared Threat to Security." *Washington Post*, April 30, 2000. https://www.washingtonpost.com/archive/politics/2000/04/30/aids-is-declared-threat-to-security/c5e976e4-3fe8-411b-9734-ca44f3130b41/.

Gibbs, E.P.J. 2005. "Emerging Zoonotic Epidemics in the Interconnected Global Community." *Veterinary Record* 157 (22): 673–679.

Goodstein, Laurie. 2001. "In U.S., Echoes of Rift of Muslims and Jews." *New York Times*, September 12, 2001, late edition, sec. A.

Government Accounting Office. 2003. "Smallpox Vaccination: Implementation of National Program Faces Challenges." No. GAO-03-578 (April). https://www.gao.gov/products/GAO-03-578.

Grabenstein, John D., Phillip R. Pittman, John T. Greenwood, and Renata J. M. Engler. 2006. "Immunization to Protect the US Armed

Forces: Heritage, Current Practice, and Prospects." *Epidemiologic Reviews* 28 (1): 3–26. https://doi.org/10.1093/epirev/mxj003.

Greene, Jeremy, Marguerite Thorp Basilico, Heidi Kim, and Paul Farmer. 2013. "Colonial Medicine and Its Legacies." In *Reimagining Global Health: An Introduction*, edited by Paul Farmer, Arthur Kleinman, Jim Kim, and Matthew Basilico, 33–73. Berkeley: University of California Press.

Greensboro News and Record. 2002. "Anthrax Search Reveals Seamier Side of Science," editorial, January 25, 2002.

Greger, Michael. 2006. *Bird Flu: A Virus of Our Own Hatching.* New York: Lantern Books.

Grewal, Inderpal. 2005. *Transnational America: Feminisms, Diasporas, Neoliberalisms.* Durham, NC: Duke University Press.

———. 2006. "'Security Moms' in the Early Twentieth-Century United States: The Gender of Security in Neoliberalism." *Women's Studies Quarterly* 34 (1/2): 25–39.

Guillemin, Jeanne. 1999. *Anthrax: The Investigation of a Deadly Outbreak.* Berkeley: University of California Press.

———. 2005a. *Biological Weapons: From the Invention of State-Sponsored Programs to Contemporary Bioterrorism.* New York: Columbia University Press.

———. 2005b. "Inventing Bioterrorism." In *Making Threats: Biofears and Environmental Anxieties*, edited by Betsy Hartmann, Banu Subramaniam, and Charles Zerner, 197–216. Lanham, MD: Rowman and Littlefield.

Hall, Stuart. 1980. "Encoding/Decoding." In *Culture, Media, Language*, edited by Stuart Hall, Dorothy Hobson, Andrew Love, and Paul Willis, 128–138. London: Hutchinson.

———. 1992. "The West and the Rest: Discourse and Power." In *Formations of Modernity*, edited by Stuart Hall and Bram Gieben. Cambridge, UK: Polity.

———. 1997. "The Spectacle of the 'Other.'" In *Representation: Cultural Representations and Signifying Practices*, 223–290. London: Sage.

Hammond, Edward. 2007. "Should the US and Russia Destroy Their Stocks of Smallpox Virus?" *BMJ: British Medical Journal* 334 (7597): 774. https://doi.org/10.1136/bmj.39155.695255.94.

————. 2009. "Some Intellectual Property Issues Related to H5N1 Influenza Viruses, Research and Vaccines." Third World Network. http://www.twn.my/title2/IPR/ipr12.htm.

Haraway, Donna. 1991. "A Cyborg Manifesto: Science, Technology, and Socialist-Feminism in the Late 20th Century." In *Simians, Cyborgs, and Women: The Reinvention of Nature*, 183–202. New York: Routledge.

Hart, John. 2006. "The Soviet Biological Weapons Program." In *Deadly Cultures: Biological Weapons since 1945*, edited by Mark Wheelis, Lajos Rózsa, and Malcolm Dando, 132–156. Boston: Harvard University Press.

Hatem, Mervat. 2011. "The Political and Cultural Representations of Arabs, Arab Americans, and Arab American Feminists after September 11, 2001." In *Arab and Arab American Feminisms: Gender, Violence, and Belonging*, edited by Rabab Abdulhadi, Evelyn Alsultany, and Nadine Naber, 10–28. Syracuse, NY: Syracuse University Press. https://muse.jhu.edu/book/12992.

Hawthorne, Susan. 2007. "Land, Bodies, and Knowledge: Biocolonialism of Plants, Indigenous Peoples, Women, and People with Disabilities." *Signs* 32 (2): 314–323. https://doi.org/10.1086/508224.

Healthcare-NOW. 2009. "AFL-CIO Convention Endorses Single-Payer—Healthcare-NOW!" September 15. https://www.healthcare-now.org/blog/afl-cio-convention-endorses-single-payer/.

Hecht, Jeff, and Debora MacKenzie. 2005. "Safety Fears Raised over Biosecurity Lapse." *New Scientist*, January 20. https://www.newscientist.com/article/dn6903-safety-fears-raised-over-biosecurity-lapse/.

Henninger, Daniel. 2003. "Wonder Land: Gangs of the World Make Barbarism Look Real Enough." *Wall Street Journal*, January 3, 2003, sec. A.

Hersh, Seymour M. 1968. *Chemical and Biological Warfare: America's Hidden Arsenal*. Indianapolis: Bobbs-Merrill.

Heymann, David L. 2004. "The International Response to the Outbreak of SARS." *Philosophical Transactions of the Royal Society of London. Series B, Biological Sciences* 359 (August): 1127–1129. https://doi.org/10.1098/rstb.2004.1484.

Heymann, David L., and Guenael Rodier. 2004. *SARS: Lessons from a New Disease*. National Academies Press. https://www.ncbi.nlm.nih.gov/books/NBK92444/.

"HHS Fact Sheet: Biodefense Preparedness." 2004. U.S. Department of
Health and Human Services. News & Policies. April 28. https://
georgewbush-whitehouse.archives.gov/news/releases/2004/04
/20040428-4.html.

"HHS Will Lead Government-Wide Effort to Enhance Biosecurity in
'Dual Use' Research." 2004. *HHS News*, March 4, 2004.

Hine, Darlene Clark. 1989. *Black Women in White: Racial Conflict and
Cooperation in the Nursing Profession, 1890–1950*. Bloomington: Indiana
University Press.

Hirsch, Robert. 2005. "The Strange Case of Steve Kurtz: Critical Art
Ensemble and the Price of Freedom." *Afterimage*, June.

Hochberg, Gil Z. 2015. *Visual Occupations: Violence and Visibility in a
Conflict Zone*. Durham, NC: Duke University Press.

Holbrooke, Richard, and Laurie Garrett. 2008. "'Sovereignty' That Risks
Global Health." *Washington Post*, August 10, 2008, Opinion sec.
http://www.washingtonpost.com/wp-dyn/content/article/2008/08/08
/AR2008080802919.html.

Homeland Security Council. 2005. *National Strategy for Pandemic
Influenza*. https://www.cdc.gov/flu/pandemic-resources/pdf/pandemic
-influenza-strategy-2005.pdf.

———. 2006. *National Strategy for Pandemic Influenza Implementation
Plan*. https://www.cdc.gov/flu/pandemic-resources/pdf/pandemic
-influenza-implementation.pdf.

hooks, bell. 1996. Introduction to *Reel to Real: Race, Sex, and Class at the
Movies*. New York: Routledge.

Huang, Yanzhong. 2004. "The SARS Epidemic and Its Aftermath in
China: A Political Perspective." In *Learning from SARS: Preparing for
the Next Disease Outbreak: Workshop Summary*. Washington, DC:
National Academies Press. https://www.ncbi.nlm.nih.gov/books
/NBK92479/.

Hunter, Cameron, and Sammy Salama. 2006. "Iraq's WMD Scientists in
the Crossfire." Nuclear Threat Initiative. http://www.nti.org/analysis
/articles/iraqs-wmd-scientists-crossfire/.

Huntington, Samuel P. 1993. "The Clash of Civilizations." *Foreign Affairs*
72 (3): 22–49.

Huxsoll, D. L., C. D. Parrott, and W. C. Patrick. 1989. "Medicine in Defense against Biological Warfare." *JAMA* 262 (5): 677–679.

Ibish, Hussein. 2003. *Report on Hate Crimes and Discrimination against Arab Americans: The Post-September 11 Backlash, September 11, 2001–October 11, 2002.* Washington, DC: American-Arab Anti-Discrimination Committee.

Indigenous Peoples Council on Biocolonialism. n.d. "IPCB Programs and Objectives." Accessed August 13, 2019. http://www.ipcb.org/about_us /our_mission.html.

Inglesby, Thomas V., Tara O'Toole, Donald A. Henderson, John G. Bartlett, Michael S. Ascher, Edward Eitzen, Arthur M. Friedlander, et al. 2002. "Anthrax as a Biological Weapon, 2002: Updated Recommendations for Management." *JAMA* 287 (17): 2236–2252.

Institute for Public Accuracy. 2003. "Weapons Inspectors Going to Work in America." February 28. http://www.accuracy.org/release/549 -weapons-inspectors-going-to-work-in-america/.

Institute of Medicine. 2003. "A Case in Point: Influenza—We Are Unprepared." In *Microbial Threats to Health: Emergence, Detection, and Response,* edited by Mark S. Smolinski, Margaret A. Hamburg, and Joshua Lederberg. Washington, DC: National Academies Press. http://www.ncbi.nlm.nih.gov/books/NBK221486/.

———. 2004. *Learning from SARS: Preparing for the Next Disease Outbreak—Workshop Summary.* Washington, DC: National Academies Press. https://www.nap.edu/catalog/10915/learning-from-sars -preparing-for-the-next-disease-outbreak-workshop.

———. 2005. "The Implementation of the Smallpox Vaccination Program." In *The Smallpox Vaccination Program: Public Health in an Age of Terrorism.* Washington, DC: National Academies Press. https://doi.org /10.17226/11240.

———. 2008. *Preparing for an Influenza Pandemic: Personal Protective Equipment for Healthcare Workers.* Washington, DC: National Academies Press. https://www.nap.edu/catalog/11980/preparing-for-an -influenza-pandemic-personal-protective-equipment-for-healthcare.

"Iraq Body Count." n.d. Accessed August 7, 2019. https://www .iraqbodycount.org/database/.

"Iraq's Jailed Mrs Anthrax 'Dying.'" 2005. *BBC News*, January 1, 2005. http://news.bbc.co.uk/2/hi/middle_east/4138767.stm.

Ismail, M. Asif. 2007. "Spending on Lobbying Thrives." Center for Public Integrity, April 1. https://www.publicintegrity.org/2007/04/01/5780 /spending-lobbying-thrives.

Itagaki, Lynn. 2016. Introduction to *Civil Racism*. Minneapolis: University of Minnesota Press.

Jamal, Amaney. 2008. "Civil Liberties and the Otherization of Arab and Muslim Americans." In *Race and Arab Americans before and after 9/11: From Invisible Citizens to Visible Subjects*, edited by Amaney Jamal and Nadine Naber, 114–130. Syracuse, NY: Syracuse University Press.

Jaschik, Scott. 2005. "Why Is Huda Ammash behind Bars?" *Inside Higher Ed*, August 25. https://www.insidehighered.com/news/2005/08/25 /ammash.

Johnson, Tim. 2006. "China Shares Some Flu Samples from Birds." *Pittsburgh Post-Gazette*, November 11. https://news.google.com /newspapers?nid=1129&dat=20061111&id=W7YNAAAAIBAJ&sjid =WHIDAAAAIBAJ&pg=4307,9449&hl=en.

Jones, James. 1993. *Bad Blood: The Tuskegee Syphilis Experiment, New and Expanded Edition*. Rev. ed. New York: Free Press.

Joseph, Suad, Benjamin D'Harlingue, and Alvin Ka Hin Wong. 2008. "Arab Americans and Muslim Americans in the *New York Times*, before and after 9/11." In *Race and Arab Americans before and after 9/11: From Invisible Citizens to Visible Subjects*, edited by Amaney Jamal and Nadine Naber, 229–275. Syracuse, NY: Syracuse University sity Press.

Kaiser, Jocelyn. 2005. "Researchers Relieved by Final Biosecurity Rules." *Science* 308 (5718): 31. https://doi.org/10.1126/science.308.5718.31a.

———. 2007. "Biosafety Breaches: Accidents Spur a Closer Look at Risks at Biodefense Labs." *Science* 317 (5846): 1852–1854. https://doi.org/10.1126 /science.317.5846.1852.

Kane, Stephanie, and Pauline Greenhill. 2007. "A Feminist Perspective on Bioterror: From Anthrax to Critical Art Ensemble." *Signs* 33 (1): 53–80. https://doi.org/10.1086/518261.

Kaufman, Joan. 2006. "SARS and China's Health-Care Response: Better to Be Both Red and Expert!" In *SARS in China: Prelude to Pandemic?*, edited by Arthur Kleinman and James L. Watson, 53–68. Stanford, CA: Stanford University Press.

Kelle, Alexander, Kathryn Nixdorff, and Malcolm Dando. 2012. *Preventing a Biochemical Arms Race*. Stanford, CA: Stanford University Press.

Khaliq, Zeeshan, Mikael Leijon, Sándor Belák, and Jan Komorowski. 2016. "Identification of Combinatorial Host-Specific Signatures with a Potential to Affect Host Adaptation in Influenza A H1N1 and H3N2 Subtypes." *BMC Genomics* 17 (July): 529. https://doi.org/10.1186/s12864 -016-2919-4.

Khan, A. S., S. Morse, and S. Lillibridge. 2000. "Public-Health Preparedness for Biological Terrorism in the USA." *Lancet* 356 (9236): 1179–1182. https://doi.org/10.1016/S0140-6736(00)02769-0.

Khor, Martin. 2007. "Developing Countries Call for New Flu Virus Sharing System." http://www.twn.my/title2/avian.flu/news.stories/afns .006.htm.

Kim, Jodi. 2010. *Ends of Empire: Asian American Critique and the Cold War*. Minneapolis: University of Minnesota Press.

Kimball, Daryl G. 2003. "Iraq's WMD: Myth and Reality." Arms Control Association, September 1. https://www.armscontrol.org/act/2003_09 /focus_Septo3.

King, Jonathan, and Harlee Strauss. 1990. "The Hazards of Defensive Biological Warfare Programs." In *Preventing a Biological Arms Race*, edited by Susan Wright, 120–132. Boston: MIT Press.

King, Nicholas B. 2002. "Security, Disease, Commerce: Ideologies of Postcolonial Global Health." *Social Studies of Science* 32 (5/6): 763–789.

———. 2003. "The Influence of Anxiety: September 11, Bioterrorism, and American Public Health." *Journal of the History of Medicine and Allied Sciences* 58 (4): 433–441.

———. 2004. "The Scale Politics of Emerging Diseases." *Osiris* 19 (1): 62–76. https://doi.org/10.1086/649394.

Kissinger, Henry A. 1969. "National Security Decision Memorandum 35." National Security Council, November 25. https://2001-2009.state.gov/r /pa/ho/frus/nixon/e2/83596.htm.

Kitler, M. E., P. Gavinio, and D. Lavanchy. 2002. "Influenza and the Work of the World Health Organization." *Vaccine* 20 Suppl 2 (May): S5-14.

Kraut, Alan. 1994. *Silent Travelers: Germs, Genes, and the "Immigrant Menace."* New York: Basic Books.

Kristof, Nicholas D. 2002. "Iraq's Little Secret." *New York Times*, October 1, 2002, Opinion. https://www.nytimes.com/2002/10/01/opinion /iraq-s-little-secret.html.

Krupnick, Matt. 2003. "CNA Calls Smallpox Program 'Political': Nursing Union Official Says Bush's Plan Carries More Risks Than Benefits." *Contra Costa Times*, January 24, 2003. https://www.mail-archive.com /ctrl@listserv.aol.com/msg102172.html.

Kuhles, Daniel, and David Ackman. 2003. "The Federal Smallpox Vaccination Program: Where Do We Go from Here?" *Health Affairs* 10 (October): 1377.

Kwik, Gigi, Joe Fitzgerald, Thomas V. Inglesby, and Tara O'Toole. 2003. "Biosecurity: Responsible Stewardship of Bioscience in an Age of Catastrophic Terrorism." *Biosecurity and Bioterrorism: Biodefense Strategy, Practice, and Science* 1 (1): 27–35. https://doi.org/10.1089 /15387130360514805.

Lakoff, Andrew. 2008a. "From Population to Vital System: National Security and the Changing Object of Public Health." In *Biosecurity Interventions: Global Health and Security in Question*, edited by Andrew Lakoff and Stephen J. Collier, 33–60. New York: Columbia University Press.

———. 2008b. "The Generic Biothreat, or, How We Became Unprepared." *Cultural Anthropology* 23 (3): 399–428. https://doi.org/10.1111/j.1548-1360 .2008.00013.x.

———. 2010. "Two Regimes of Global Health." *Humanity: An International Journal of Human Rights, Humanitarianism, and Development* 1 (1): 59–79. https://doi.org/10.1353/hum.2010.0001.

———. 2015. "Global Health Security and the Pathogenic Imaginary." In *Dreamscapes of Modernity: Sociotechnical Imaginaries and the Fabrication of Power*, edited by Sheila Jasanoff and Sang-Hyun Kim, 300–320. Chicago: University of Chicago Press. https://www.researchgate.net /publication/281114115_Global_Health_Security_and_the_Pathogenic _Imaginary.

Lakoff, Andrew, and Stephen J. Collier, eds. 2008. *Biosecurity Interventions: Global Health and Security in Question*. New York: Columbia University Press.

Landesman, Peter. 2002. "The Year in Ideas; Smallpox Martyrs." *New York Times*, December 15, 2002.

Landro, Laura. 2002. "The Informed Patient: Don't Leave It All to Doctors to Know Signs of Bioweapons." *Wall Street Journal*, September 12, 2002, Eastern edition.

Lashley, Felissa R., and Jerry D. Durham. 2002. *Emerging Infectious Diseases: Trends and Issues*. New York: Springer.

Leavitt, Judith Walzer. 1996. *Typhoid Mary: Captive to the Public's Health*. Boston: Beacon Press.

Lederberg, Joshua. 1968. "Congress Should Examine Biological Warfare Tests." *Washington Post*, March 30, 1968.

Lederberg, Joshua, Robert E. Shope, and Stanley C. Oaks. 1992. *Emerging Infections: Microbial Threats to Health in the United States*. https://www.nap.edu/catalog/2008/emerging-infections-microbial-threats-to-health-in-the-united-states.

Lee, Charles T. 2009. "Suicide Bombing as Acts of Deathly Citizenship? A Critical Double-Layered Inquiry." *Critical Studies on Terrorism* 2 (2): 147–163. https://doi.org/10.1080/17539150903010236.

Lee, Erika. 2007. "The 'Yellow Peril' and Asian Exclusion in the Americas." *Pacific Historical Review* 76 (4): 537–562. https://doi.org/10.1525/phr.2007.76.4.537.

Lee, Grace, and Malcolm Warner. 2008. *The Political Economy of the SARS Epidemic: The Impact on Human Resources in East Asia*. London: Routledge.

Leinwand, Donna, and Laura Parker. 2003. "Troops Raid House of 'Dr. Germ'; U.S. Says Iraqi Microbiologist Led Lab Weaponizing Anthrax." *USA Today*, April 17, 2003, first edition, sec. A.

Levich, Jacob. 2015. "The Gates Foundation, Ebola, and Global Health Imperialism." *American Journal of Economics and Sociology* 74 (4): 704–742. https://doi.org/10.1111/ajes.12110.

Lin, Cindy. 2015. "Examining International and Indonesian Responses to H5N1 Influenza." *Subjectivities: A Journal of Perspectives on Southeast Asia*, July. https://sbjtvt.wordpress.com/2015/07/18/examining-international-and-indonesian-responses-to-h5n1-influenza-cindy-lin/.

Loomba, Ania. 2015. *Colonialism/Postcolonialism*. 3rd ed. London: Routledge.

Los Angeles Times. 2004. "Commentary: Muslims and Assimilation." April 24, 2004, Home edition.

Lowe, Lisa. 1996. "Immigration, Citizenship, Racialization: Asian American Critique." In *Immigrant Acts: On Asian American Cultural Politics*, 1–36. Durham, NC: Duke University Press.

Lupton, Deborah. 1995. *The Imperative of Health: Public Health and the Regulated Body*. Thousand Oaks, CA: Sage.

Lynch, David J. 2003. "Wild Animal Markets May Be Breeding SARS; In Southern China, Rats, Snakes, Cats and Dogs Still Sold as Food despite Likely Link to Virus." *USA Today*, October 29, 2003.

Machi, Vivienne. 2017. "Homeland Security Struggling to Fund Chem-Bio Defense." *National Defense*, September 22, 2017. http://www.nationaldefensemagazine.org/articles/2017/9/22/homeland-security-struggling-to-fund-chem-bio-defense.

MacKenzie, Debora. 2007. "Plague of Bioweapons Accidents Afflicts the US." *New Scientist*, July 5, 2007. https://www.newscientist.com/article/dn12197-plague-of-bioweapons-accidents-afflicts-the-us/.

Macleod, Roy. 1993. "Passages in Imperial Science: From Empire to Commonwealth." *Journal of World History* 4 (1): 117–150.

Mahy, Brian W. J., Jeffrey W. Almond, Kenneth I. Berns, Robert M. Chanock, Dmitry K. Lvov, Ralf F. Pettersson, Hermann G. Schatzmayr, and Frank Fenner. 1993. "The Remaining Stocks of Smallpox Virus Should Be Destroyed." *Science* 262 (5137): 1223–1224.

Maira, Sunaina. 2009. "'Good' and 'Bad' Muslim Citizens: Feminists, Terrorists, and US Orientalisms." *Feminist Studies* 35 (3): 631–656.

Majaj, Lisa Suhair. 1999. "Arab-American Ethnicity: Locations, Coalitions, and Cultural Negotiations." In *Arabs in America: Building a New Future*, edited by Michael W. Suleiman, 320–336. Philadelphia: Temple University Press.

Malakoff, David. 2004. "Biosecurity Goes Global." *Science* 305 (5691): 1706–1707.

Mamdani, Mahmood. 2004a. "Afghanistan: The High Point in the Cold War." In *Good Muslim, Bad Muslim: America, the Cold War, and the Roots of Terror*, 119–177. New York: Pantheon Books.

———. 2004b. "Culture Talk: Or, How Not to Talk about Islam and Politics." In *Good Muslim, Bad Muslim: America, the Cold War, and the Roots of Terror*, 17–62. New York: Pantheon Books.

———. 2004c. "From Proxy War to Open Aggression." In *Good Muslim, Bad Muslim: America, the Cold War, and the Roots of Terror,* 178–228. New York: Pantheon Books.

Manning, Anita. 2005. "The 'Perfect Incubator'; Bird Flu Puts Health Officials on Alert for Pandemic." *USA Today,* February 8, 2005.

Marcus, George. 1995. "Ethnography in/of the World System: The Emergence of Multi-sited Ethnography." *Annual Review of Anthropology* 24: 95–117.

Markon, Jerry. 2003. "Activists' 'Inspection' Ends at Base's Gate." *Washington Post,* February 24, 2003, Metro sec.

Marks, Shula. 1997. "What Is Colonial about Colonial Medicine? And What Has Happened to Imperialism and Health?" *Social History of Medicine* 10: 205–219.

Márquez, John D., and Junaid Rana. 2015. "Introduction: On Our Genesis and Future." *Critical Ethnic Studies* 1 (1): 6.

Martin, Emily. 1994. *Flexible Bodies: Tracing Immunity in American Culture from the Days of Polio to the Age of AIDS.* Boston: Beacon Press.

Martin, William. 2004. "Legal and Public Policy Responses of States to Bioterrorism." *American Journal of Public Health* 94 (7): 1093–1096. https://doi.org/10.2105/AJPH.94.7.1093.

Masco, Joseph. 1999. "States of Insecurity: Plutonium and Post-Cold War Anxiety in New Mexico, 1992–96." In *Cultures of Insecurity: States, Communities, and the Production of Danger,* edited by Jutta Weldes, Mark Laffey, Hugh Gusterson, and Raymond Duvall, 203–232. Minneapolis: University of Minnesota Press.

Massumi, Brian. 2010. "The Future Birth of the Affective Fact: The Political Ontology of Threat." In *The Affect Theory Reader,* edited by Melissa Gregg and Gregory J. Seigworth, 52–70. Durham, NC: Duke University Press.

McCombs, Phil. 2003. "The Fire This Time: To Some Scholars, Iraq's Just Part of Something Bigger." *Washington Post,* April 13, 2003, final edition, Style sec.

McGlinchey, David. 2004. "Top U.S. Health Official Says 'Vast Majority' of States Ready for Smallpox." *Global Security Newswire,* January 30. http://www.nti.org/gsn/article/top-us-health-official-says-vast-majority-of-states-ready-for-smallpox/.

McNeil, Donald G. 2007. "Scientists Hope Vigilance Stymies Avian Flu Mutations." *New York Times*, March 27, 2007, sec. F.1.

Melosh, Barbara. 1982. *"The Physician's Hand": Work Culture and Conflict in American Nursing.* Philadelphia: Temple University Press.

Miller, Judith, William J. Broad, and Stephen Engelberg. 2001. *Germs: Biological Weapons and America's Secret War.* Reprint. New York: Simon and Schuster.

Miller, Judith, Stephen Engelberg, and William J. Broad. 2001. "U.S. Germ Warfare Research Pushes Treaty Limits." *New York Times*, September 4, 2001. https://www.nytimes.com/2001/09/04/world/us -germ-warfare-research-pushes-treaty-limits.html.

Mintz, John. 2004. "Technical Hurdles Separate Terrorists from Biowarfare." *Washington Post*, December 30, 2004, Final edition, sec. A.

Moallem, Minoo. 2002. "Whose Fundamentalism?" *Meridians* 2 (2): 298–301.

Mohanty, Chandra Talpade. 1984. "Under Western Eyes: Feminist Scholarship and Colonial Discourses." *Boundary 2* 12 (3): 333–358.

Molina, Natalia. 2006. *Fit to Be Citizens? Public Health and Race in Los Angeles, 1879–1939.* Berkeley: University of California Press.

Mueller, John. 2005. "Simplicity and Spook: Terrorism and the Dynamics of Threat Exaggeration." *International Studies Perspectives* 6 (2): 208–234. https://doi.org/10.1111/j.1528-3577.2005.00203.x.

Mukhopadhyay, Arun G. 2013. "Public Health, Genomics and Biopolitics—Human Security vis-à-vis Securing Exception." *African Journal of Political Science and International Relations* 7 (3): 133–141. https://doi.org/10.5897/AJPSIR09.077.

Muscati, Sina Ali. 2002. "Arab/Muslim 'Otherness': The Role of Racial Constructions in the Gulf War and the Continuing Crisis with Iraq." *Journal of Muslim Minority Affairs* 22 (1): 131–148.

Naber, Nadine. 2000. "Ambiguous Insiders: An Investigation of Arab American Invisibility." *Ethnic and Racial Studies* 23 (1): 37–61. https:// doi.org/10.1080/014198700329123.

———. 2008. "Look Mohammed the Terrorist Is Coming!" In *Race and Arab Americans before and after 9/11: From Invisible Citizens to Visible Subjects*, edited by Amaney Jamal and Nadine Naber, 276–304. Syracuse, NY: Syracuse University Press.

———. 2011. "Decolonizing Culture: Beyond Orientalist and Anti-
Orientalist Feminisms." In *Arab and Arab American Feminisms: Gender,
Violence, and Belonging*, edited by Rabab Abdulhadi, Evelyn Alsultany,
and Nadine Naber, 78–90. Syracuse, NY: Syracuse University Press.
http://www.jstor.org/stable/j.ctt1j1vzxf.

National Nurses United. 2012. "Medicare for All." National Nurses
United, May 23. https://www.nationalnursesunited.org/medicare
-for-all.

———. 2014. "Environmental Justice." National Nurses United, July 31.
https://www.nationalnursesunited.org/environmental-justice.

———. 2019a. "Nurses Answer the Call: RNRN/NNU Sends More
Volunteers to Care for Migrants and Asylum Seekers at the Border."
National Nurses United, March 1. https://www.nationalnursesunited
.org/press/nurses-answer-call-rnrnnnu-sends-more-volunteers-care
-migrants-and-asylum-seekers-border.

———. 2019b. "RNs of National Nurses United Say Racism, Xenophobia
Combined with Lax Gun Control Laws at Root of Mass Shooting
Epidemic." National Nurses United, August 4. https://www
.nationalnursesunited.org/press/rns-national-nurses-united-say-racism
-xenophobia-combined-lax-gun-control-laws-root-mass.

*The National Pandemic Influenza Preparedness and Response Plan: Is the
United States Ready for Avian Flu?: Hearing before the Committee on
Government Reform.* 2005. 109th Cong., 1st sess. (November 4). https://
www.govinfo.gov/content/pkg/CHRG-109hhrg24820/html/CHRG
-109hhrg24820.htm.

National Research Council. 2004. *Biotechnology Research in an Age of
Terrorism.* https://doi.org/10.17226/10827.

———. 2011. *Review of the Scientific Approaches Used during the FBI's
Investigation of the 2001 Anthrax Letters.* https://doi.org/10.17226/13098.

National Security Council. 2009. *National Strategy for Countering Biological
Threats.* November 23. https://www.hsdl.org/?view&did=31404.

Nayak, Meghana. 2006. "Orientalism and 'Saving' US State Identity after
9/11." *International Feminist Journal of Politics* 8 (1): 42–61.

Nelson, Alondra. 2013. *Body and Soul: The Black Panther Party and the Fight
against Medical Discrimination.* Minneapolis: University of Minnesota
Press.

Newkirk, Vann R. II. 2017. "How Trump's Budget Would Weaken Public Health." *The Atlantic*, May 23, 2017. https://www.theatlantic.com/politics/archive/2017/05/trump-budget-public-health/527808/.

Newman, Maria, and Yilu Zhao. 2003. "Fear, Not SARS, Rattles South Jersey School." *New York Times*, May 10, 2003, sec. B.

NewsBank. n.d. "Access World News." Accessed March 28, 2008. https://www.newsbank.com/libraries/colleges-universities/solutions/top-resources/access-world-news.

Nonproliferation Studies, James Martin Center for. 2004. "Limiting the Acquisition and Use of Biological Weapons by Strengthening the BWC." Monterey Institute of International Studies. WMD 411: Your Information Resource on Nuclear, Biological and Chemical Weapons Issues.

Nuclear Threat Initiative. n.d. "Iraq Profile: Biological Overview." Accessed December 1, 2006. https://www.nti.org/learn/countries/iraq/biological/.

———. 2001. "Smallpox: CDC Trains Health Officials for Smallpox Outbreak." Global Security Newswire, December 18.

———. 2002. "Iraq: Russian Scientist Might Have Delivered Potent Smallpox Strain." National Journal Group. Global Security Newswire, December 3.

———. 2010. "Pentagon Pulls $1B from WMD-Defense Efforts to Fund Vaccine Initiative." Global Security Newswire, August 27, 2010. https://www.nti.org/gsn/article/pentagon-pulls-1b-from-wmd-defense-efforts-to-fund-vaccine-initiative/.

O'Neill, Peter D. 2017. "Introduction: Famine Irish and the American Racial State." In *Famine Irish and the American Racial State*, 1–31. New York: Routledge.

Ong, Aihwa. 2008. "Scales of Exception: Experiments with Knowledge and Sheer Life in Tropical Southeast Asia." *Singapore Journal of Tropical Geography* 29 (2): 117–129. https://doi.org/10.1111/j.1467-9493.2008.00323.x.

OPCW. 2019. "Engineering Assessment of Two Cylinders Observed at the Douma Incident." http://syriapropagandamedia.org/working-papers/assessment-by-the-engineering-sub-team-of-the-opcw-fact-finding

-mission-investigating-the-alleged-chemical-attack-in-douma-in-april
-2018.

O'Toole, Tara. 2001. "Emerging Illness and Bioterrorism: Implications for
Public Health." *Journal of Urban Health: Bulletin of the New York Academy
of Medicine* 78 (2): 396–402. https://doi.org/10.1093/jurban/78.2.396.

O'Toole, Tara, and Thomas V. Inglesby. 2004. "Toward Biosecurity."
Biosecurity and Bioterrorism: Biodefense Strategy, Practice, and Science 1
(1): 1–4. https://doi.org/10.1089/15387130360514760.

O'Toole, Tara, Mair Michael, and Thomas V. Inglesby. 2002. "Shining
Light on 'Dark Winter.'" *Clinical Infectious Diseases* 34 (7): 972–983.
https://doi.org/10.1086/339909.

Pazola, Ron. 2001. "County Unveils Bioterrorism Plan." *Naperville Sun*,
November 28, 2001, sec. 9.

Pearson, Helen, Tom Clarke, Alison Abbott, Jonathan Knight, and David
Cyranoski. 2003. "SARS: What Have We Learned?" *Nature* 424
(6945): 121–126.

People's World. 2003. "Inspectors Say: Open U.S. Weapons Sites, Too,"
February 28, 2003. https://www.peoplesworld.org/article/inspectors
-say-open-u-s-weapons-sites-too/.

Perez-Pena, Richard. 2003. "Checking City's Archives to Solve a Medical
Mystery." *New York Times*, October 3, 2003, late edition, sec. B.

Philip, Kavita. 2004. *Civilizing Natures: Race, Resources, and Modernity in
Colonial South India*. New Brunswick, NJ: Rutgers University Press.

———. 2015. "Telling Histories of the Future: The Imaginaries of Indian
Technoscience." *Identities* 23 (April): 1–18.

Philipose, Elizabeth. 2008. "Decolonizing the Racial Grammar of
International Law." In *Feminism and War: Confronting US Imperialism*,
1st ed., edited by Robin L. Riley, Chandra Talpade Mohanty, and
Minnie Bruce Pratt, 103–16. London: Zed Books.

Piller, Charles, and Keith R. Yamamoto. 1990. "The U.S. Biological
Defense Research Program in the 1980s." In *Preventing a Biological
Arms Race*, edited by Susan Wright, 133–168. Boston: MIT Press.

Pitt, William. 2002. *War on Iraq: What Team Bush Doesn't Want You to
Know*. New York: Context Books.

Poster, Mark. 1995. *The Second Media Age*. Cambridge, UK: Polity.

Powell, Colin. 2003. "Secretary of State Addresses the U.N. Security Council." February 5. https://georgewbush-whitehouse.archives.gov /news/releases/2003/02/print/20030205-1.html#.

Prashad, Vijay. 2007. "Paris." In *The Darker Nations: A People's History of the Third World*, 3–15. New York: New Press.

———. 2017. "The Time of the Popular Front." *Third World Quarterly* 38 (11): 2536–2545. https://doi.org/10.1080/01436597.2017.1350103.

Puar, Jasbir. 2007a. "The Sexuality of Terrorism." In *Terrorist Assemblages: Homonationalism in Queer Times*, 37–78. Durham, NC: Duke University Press.

———. 2007b. "'The Turban Is Not a Hat': Queer Diaspora and Practices of Profiling." In *Terrorist Assemblages: Homonationalism in Queer Times*, 166–202. Durham, NC: Duke University Press.

Puar, Jasbir, and Amit S. Rai. 2002. "Monster, Terrorist, Fag: The War on Terrorism and the Production of Docile Patriots." *Social Text* 20 (3): 117–148.

Public Health Security and Bioterrorism Preparedness and Response Act of 2002. 2002. 116 Stat. 594. https://www.govtrack.us/congress/bills/107 /hr3448.

Quigley, J. 1992. "The Legality of the Biological Defense Research Program." *Annals of the New York Academy of Sciences* 666 (December): 131–145.

Ramirez, Michael. 2001. *Los Angeles Times*, October 13, 2001, sec. B.

Rana, Junaid. 2013. "Tracing the Muslim Body: Race, U.S. Deportation, and Pakistani Return Migration." In *The Sun Never Sets: South Asian Migrants in the Age of U.S. Power*, edited by Vivek Bald, Miabi Chatterji, Sujani Reddy, and Manu Vimalassary, 325–349. New York: New York University Press. https://experts.illinois.edu/en/publications /tracing-the-muslim-body-race-us-deportation-and-pakistani-return-.

RAND. 2003. "Widespread Smallpox Vaccination Is Too Dangerous." https://www.rand.org/news/pox.html.

Rapoport, David. 1999. "Terrorism and Weapons of the Apocalypse." *National Security Studies Quarterly* 5 (1): 49–66. https://www .researchgate.net/publication/242769630_Terrorism_and_Weapons_of _the_Apocalypse.

Rasenberger, Jim. 2003. "A City in the Time of Scourge: As the Specter of Smallpox Rises Again, an Old Triumph Offers a Little Encouragement." *New York Times*, April 6, 2003, City Lore edition, sec. CY.

Respiratory Diseases Committee of the American Association of Avian Pathologists. n.d. "Asian Bird Flu." USDA Agricultural Research Service. Accessed December 1, 2005. https://www.ars.usda.gov /southeast-area/athens-ga/us-national-poultry-research-center/exotic -emerging-avian-viral-diseases-research/docs/asian-bird-flu/.

Reverby, Susan M. 1987. *Ordered to Care: The Dilemma of American Nursing, 1850–1945.* Cambridge, UK: Cambridge University Press.

Ridgeway, James. 2005. "Capitalizing on the Flu." *Village Voice*, November 15, News and Politics sec. https://www.villagevoice.com/2005/11/15 /capitalizing-on-the-flu/.

Riley, Robin L. 2004. "Huda, Rihab and the Missing Iraqi Woman: Orientalism and the New Sexism in Representations of the War on Iraq." Presented at the International Studies Association, Montreal, Quebec, Canada, March 17.

Riley, Robin L., Chandra Talpade Mohanty, and Minnie Bruce Pratt, eds. 2008. *Feminism and War: Confronting US Imperialism.* 1st ed. London: Zed Books.

Robertson, Tatsha. 2003. "Antiwar Protesters Try New Tactics: Activists to Flood Political Leaders with Calls, E-Mails." *Boston Globe*, February 24, 2003.

Rose, Nikolas. 1999. *Powers of Freedom: Reframing Political Thought.* Cambridge, UK: Cambridge University Press.

Rosenbaum, Eli. M. 1998. "Re: U.S. Non-prosecution of Japanese War Criminals." Department of Justice, Office of Special Investigations. December 17. https://assets.documentcloud.org/documents/3720697 /DOJ-Copy-Cooper-1998-Correspondence.pdf.

Rosenberg, Barbara Hatch. 2003. "Secret Biodefense Activities Are Undermining the Norm against Biological Weapons." Federation of American Scientists Working Group on Biological Weapons. http:// armscontrolcenter.org/wp-content/uploads/2016/02/biodefense-FAS -position-paper.pdf.

Rosenberg, Eric. 2003. "'Arms Inspectors' to Eyeball America Canadian Groupplanning Spoof." *San Antonio Express-News*, February 9, 2003, Metro Edition, sec. A.

Rowe, Aimee Marie Carrillo. 2004. "Whose 'America'? The Politics of Rhetoric and Space in the Formation of U.S. Nationalism." *Radical History Review* 89 (1): 115–134.

Rubin, Elaine R., Lisa M. Lindeman, and Marian Osterweis, eds. 2002. *Emergency Preparedness: Bioterrorism and Beyond*. Washington, DC: Association of Academic Health Centers.

Saez, Catherine. 2018. "Shared Indigenous Knowledge and Benefit-Sharing Needs Particular Attention, Panel Tells CBD." *Intellectual Property Watch* (blog), November 29. https://www.ip-watch.org/2018/11 /29/shared-indigenous-knowledge-benefit-sharing-needs-particular -attention-panel-tells-cbd/.

Said, Edward W. 1978. *Orientalism*. London: Routledge and Kegan Paul.

Samhan, Helen Hatab. 1999. "Not Quite White: Race Classification and the Arab-American Experience." In *Arabs in America: Building a New Future*, edited by Michael W. Suleiman, 209–226. Philadelphia: Temple University Press.

Sandler, Todd. 2003. "Collective Action and Transnational Terrorism." *World Economy* 26 (6): 779–802. https://doi.org/10.1111/1467-9701.00548.

Sarasin, Philipp. 2006. *Anthrax: Bioterror as Fact and Fantasy*. Cambridge, MA: Harvard University Press.

SARS: Assessment, Outlook, and Lessons Learned: Hearing before the Subcommittee on Oversight and Investigations. 2003. 108th Cong., 1st sess. (May 7). https://www.gpo.gov/fdsys/pkg/CHRG-108hhrg87484/html /CHRG-108hhrg87484.htm.

Scheer, Robert. 2005. "Dr. Germ and Mrs. Anthrax Set Free." *Truthdig*, December 28. https://www.truthdig.com/articles/dr-germ-and-mrs -anthrax-set-free/.

Schnur, Alan. 2006. "The Role of the World Health Organization in Combating SARS, Focusing on the Efforts in China." In *SARS in China: Prelude to Pandemic?*, edited by Arthur Kleinman and James L. Watson, 31–52. Stanford, CA: Stanford University Press.

Schrijver, Remco S., and G. Koch, eds. 2005. *Avian Influenza: Prevention and Control*. Dordrect: Springer.

Science for the People. n.d.a. "About SftP." Accessed August 17, 2019. https://scienceforthepeople.org/about-sftp/.

———. n.d.b. "Bringing a Radical Perspective to the March for Science." Accessed August 17, 2019. https://scienceforthepeople.org/bringing-a-radical-perspective-to-the-march-for-science/.

———. n.d.c. "No Tech for ICE!" Accessed August 17, 2019. https://scienceforthepeople.org/no-tech-for-ice/.

———. n.d.d. "People's Green New Deal—Topical Focus Areas." Accessed August 17, 2019. https://scienceforthepeople.org/peoples-green-new-deal/focus-areas/.

———. n.d.e. "Working Groups." Accessed August 19, 2019. https://scienceforthepeople.org/working-groups/.

———. 2018. "The Dual Nature of Science." April 12. https://scienceforthepeople.org/2018/04/12/dual-nature-of-science/.

Scott, Dylan. 2018. "Trump's Health Care Budget, Explained." *Vox*, February 12. https://www.vox.com/policy-and-politics/2018/2/12/17005294/trump-health-care-budget-explained.

Sedyaningsih, Endang R., Siti Isfandari, Triono Soendoro, and Siti Fadilah Supari. 2008. "Towards Mutual Trust, Transparency and Equity in Virus Sharing Mechanism: The Avian Influenza Case of Indonesia." *Annals of the Academy of Medicine, Singapore* 37 (6): 482–488.

Shah, Nayan. 1999. "Cleansing Motherhood: Hygiene and the Culture of Domesticity in San Francisco's 'Chinatown,' 1875–1939." In *Gender, Sexuality and Colonial Modernities*, edited by Antoinette Burton, 19–34. New York: Routledge.

———. 2001. *Contagious Divides*. Berkeley: University of California Press.

Shaheen, Jack G. 1984. *The TV Arab*. Madison, WI: Popular Press.

Sharp, Phillip A. 2005. "Editorial: 1918 Flu and Responsible Science." *Science* 310 (5745): 17.

Shea, Dana E., and Sarah A. Lister. 2003. *The BioWatch Program: Detection of Bioterrorism*. Report No. 32152. Washington, DC: Library of Congress. https://digital.library.unt.edu/ark:/67531/metacrs8189/.

Shepherd, L. 2006. "Veiled References: Constructions of Gender in the Bush Administration Discourse on the Attacks on Afghanistan Post-9/11." *International Feminist Journal of Politics* 8 (1): 19–41.

Shih, Shu-mei. 2001. "Introduction: The Global and Local Terms of Chinese Modernism." In *The Lure of the Modern: Writing Modernism in Semicolonial China, 1917–1937*, 1–47. Berkeley: University of California Press.

———. 2008. "Comparative Racialization: An Introduction." *PMLA* 123 (5): 1347–1362.

Shiva, Vandana. 1997. *Biopiracy: The Plunder of Nature and Knowledge*. Boston: South End Press.

Shohat, Ella. 1991. "Gender and Culture of Empire: Toward a Feminist Ethnography of the Cinema." *Quarterly Review of Film and Video* 13 (1–3): 45–84.

Shope, Robert E., and Alfred S. Evans. 1993. "Assessing Geographic and Transport Factors, and Recognition of New Viruses." In *Emerging Viruses*, edited by Stephen S. Morse, 109–119. New York: Oxford University Press.

Shreeve, Jamie. 2006. "Why Revive a Deadly Flu Virus?" *New York Times Magazine*, January 29, 2006. https://www.nytimes.com/2006/01/29/magazine/why-revive-a-deadly-flu-virus.html.

Shrestha, Sundar, David Swerdlow, Rebekah Borse, Vimalanand S. Prabhu, Lyn Finelli, Charisma Atkins, Kwame Owusu-Edusei, et al. 2011. "Estimating the Burden of 2009 Pandemic Influenza A (H1N1) in the United States (April 2009–April 2010)." *Clinical Infectious Diseases: An Official Publication of the Infectious Diseases Society of America* 52 Suppl 1 (January): S75–82. https://doi.org/10.1093/cid/ciq012.

Shyrock, Andrew. 2008. "The Moral Analogies of Race." In *Race and Arab Americans before and after 9/11: From Invisible Citizens to Visible Subjects*, edited by Amaney Jamal and Nadine Naber, 81–113. Syracuse, NY: Syracuse University Press.

Sidel, Victor W., Robert M. Gould, and Hillel W. Cohen. 2002. "Bioterrorism Preparedness: Cooptation of Public Health?" *Medicine and Global Survival* 7: 82–89.

Singer, Merrill. 2009. "Pathogens Gone Wild? Medical Anthropology and the 'Swine Flu' Pandemic." *Medical Anthropology* 28 (July): 199–206. https://doi.org/10.1080/01459740903070451.

Sipress, Alan. 2005. "China Has Not Shared Crucial Data on Bird Flu Outbreaks, Officials Say." *Washington Post*, July 19, 2005. http://www

.washingtonpost.com/wp-dyn/content/article/2005/07/18
/AR2005071801584.html.

Smallman, Shawn. 2013. "Biopiracy and Vaccines: Indonesia and the
World Health Organization's New Pandemic Influenza Plan." *Journal
of International and Global Studies* 4 (2): 19–36.

*The Smallpox Vaccination Plan: Challenges and Next Steps: Hearing before the
Committee on Health, Education, Labor, and Pensions.* 2003. 108th Cong.,
1st sess. (January 30). https://www.govinfo.gov/content/pkg/CHRG
-108shrg84743/html/CHRG-108shrg84743.htm.

Smith, Linda Tuhiwai. 2002. *Decolonizing Methodologies—Research and
Indigenous Peoples.* London: Zed Books.

Song, Nina. 2007. "The Framing of China's Bird Flu Epidemic by U.S.
Newspapers Influential in China: How the *New York Times* and
Washington Post Linked the Image of the Nation to the Handling of
the Disease." Master's thesis. Georgia State University. http://
scholarworks.gsu.edu/communication_theses/27/.

Spinoza, Abu. 2003. "The Detention of Dr. Huda Ammash." Counter-
punch, May 7. https://www.counterpunch.org/2003/05/07/the
-detention-of-dr-huda-ammash/.

St. John, Ronald K., Arlene King, Dick de Jong, Margaret Bodie-Collins,
Susan G. Squires, and Theresa W. S. Tam. 2005. "Border Screening for
SARS." *Emerging Infectious Diseases* 11 (1): 6–10. https://doi.org/10.3201
/eid1101.040835.

Steen, Shannon. 2010. "Coda: The Black Face of US Imperialism." In
*Racial Geometries of the Black Atlantic, Asian Pacific and American
Theatre*, by Shannon Steen, 164–168. London: Palgrave Macmillan UK.
https://doi.org/10.1057/9780230297401_6.

Stern, Alexandra. 1999. "Buildings, Boundaries, and Blood: Medicaliza-
tion and Nation-Building on the U.S.-Mexico Border, 1910–1930."
Hispanic American Historical Review 79 (7): 41–81.

Steuter, Erin, and Deborah Wills. 2009. *At War with Metaphor: Media,
Propaganda, and Racism in the War on Terror.* Washington, DC:
Lexington Books.

Stewart, Colin, Marc Lavelle, and Adam Kowaltzke. 2001. *Media and
Meaning: An Introduction.* 1st ed. London: British Film Institute.

Stolberg, Sheryl Gay. 2001. "U.S. Seeks to Stock Smallpox Vaccine for Whole Nation." *New York Times*, October 18, 2001, late edition, sec. A.

Sturm, Daniel. 2003. "Sacred Inspectors and America's Weapons of Mass Destruction." *City Pulse* (Lansing, MI), April 2, 2003.

Subramaniam, Banu. 2001. "The Aliens Have Landed! Reflections on the Rhetoric of Biological Invasions." *Meridians: Feminism, Race, Transnationalism* 2 (1): 26–40.

Sung, Yun-Wing, and Fanny M. Cheung. 2003. "Catching SARS in the HKSAR: Fallout on Economy and Community." In *The New Global Threat: Severe Acute Respiratory Syndrome and Its Impacts*, edited by Tommy Koh, Aileen Plant, and Eng Hin Lee, 147–164. Singapore: World Scientific Publishing.

Sunshine Project. 2003. "Biosafety, Biosecurity, and Bioweapons: Three Agreements on Biotechnology, Health, and the Environment, and Their Potential Contribution to Biological Weapons Control." Third World Network. https://www.twn.my/title/biosecurity.htm.

———. 2004. "Map of the US Biodefense Program: High Containment Labs and Other Facilities." November 4.

———. 2005. "Some Statistics about the US Biodefense Program and Public Health." http://web.archive.org/web/20100716130420 /http://www.sunshine-project.org/biodefense/niaidfunding.html.

———. 2006. "Protection or Proliferation? High Containment and Other Facilities of the US Biodefense Program." February 20. http://web .archive.org/web/20100716132010/http://www.sunshine-project.org /biodefense/.

Supari, Siti Fadilah. 2008. *It's Time for the World to Change: Divine Hand behind Avian Influenza*. Jakarta: PT Sulaksana Watinsa.

Swanson, Stevenson. 2003. "Iraq's 'Mrs. Anthrax' Is Key Figure in Weapons Program." *Chicago Tribune*, April 11, 2003. http://articles .chicagotribune.com/2003-04-11/news/0304110247_1_biological -weapons-mrs-anthrax-baath-party.

Thacker, Eugene. 2006. "Biocolonialism, Genomics, and the Databasing of the Population." In *The Global Genome: Biotechnology, Politics, and Culture*, 133–172. Boston: MIT Press.

Third World Network. 2007a. "Joint NGO Statement on Influenza Virus
Sharing." November 15. https://www.twn.my/announcement/Joint
.NGO.Statement.on.Influenza.Virus.Sharing.htm.

———. 2007b. "WHO to Ban Genetic Engineering of Smallpox Virus."
Biosafety Information Centre. May 21. https://biosafety-info.net
/articles/biosafety-science/emerging-trends-techniques/who-to-ban
-genetic-engineering-of-smallpox-virus/.

———. 2019. "WHO: Nagoya Protocol Decision Asks WHO to Report
on Current Pathogen-Sharing Modalities." June 12. https://www.twn
.my/title2/health.info/2019/hi190602.htm.

Thompson, Marilyn W. 2003. *The Killer Strain: Anthrax and a Government
Exposed*. New York: HarperCollins.

Tomes, N. 2000. "The Making of a Germ Panic, Then and Now." *American
Journal of Public Health* 90 (2): 191–198.

Tucker, Jonathan B. 2004. "Biological Threat Assessment: Is the Cure
Worse Than the Disease?" *Arms Control Today*, October. https://www
.armscontrol.org/act/2004_10/Tucker.

———. 2006. "Avoiding the Biological Security Dilemma: A Response to
Petro and Carus." *Biosecurity and Bioterrorism: Biodefense Strategy,
Practice, and Science* 4 (2): 195–199; discussion 200–203. https://doi.org
/10.1089/bsp.2006.4.195.

———. 2010. "Seeking Biosecurity without Verification: The New U.S.
Strategy on Biothreats." Arms Control Association. https://www
.armscontrol.org/act/2010-01/seeking-biosecurity-without-verification
-new-us-strategy-biothreats

Tucker, Jonathan B., and Erin R. Mahan. 2009. "President Nixon's
Decision to Renounce the U.S. Offensive Biological Weapons
Program." Washington, DC: National Defense University Center for
the Study of Weapons of Mass Destruction. http://www.dtic.mil/docs
/citations/ADA517679.

Tung, Larry. 2003. "SARS Hits Chinatown." *Gotham Gazette*, May,
Citizen edition. https://www.gothamgazette.com/citizen/may03
/orignal_sars.shtml.

Tuohy, Lynne, and Jack Dolan. 2001. "Turmoil in a Perilous Place." *Hartford
Courant*, December 19. http://www.courant.com/hc-ant-1-story.html.

Uniting and Strengthening America by Providing Appropriate Tools Required to Intercept and Obstruct Terrorism (USA PATRIOT ACT) Act of 2001, Pub. L. No. 107–56, 115 Stat. 272 (2001). https:// www.govtrack.us/congress/bills/107/hr3162.

UNSCOM. 1999. *Status of Verification of Iraq's Biological Warfare Program*. Report to the Security Council. https://www.un.org/Depts/unscom /s99-94.htm.

USAMRIID. 2004. *USAMRIID's Medical Management of Biological Casualties Handbook*. 5th ed. Frederick, MD: USAMRIID Operational Medicine Department.

"Vaccinate against War, Not Smallpox." 2003. *Massachusetts Nurse* 74 (1): 16.

Vanzi, Max. 2004. "The Patriot Act, Other Post-9/11 Enforcement Powers and the Impact on California's Muslim Communities." Collingdale, PA: Diane Publishing.

Vaughan, Meghan. 1991. *Curing Their Ills: Colonial Power and African Illness*. Stanford, CA: Stanford University Press.

Vezzani, Simone. 2010. "Preliminary Remarks on the Envisaged World Health Organization Pandemic Influenza Preparedness Framework for the Sharing of Viruses and Access to Vaccines and Other Benefits." *Journal of World Intellectual Property* 13 (6). https://doi.org/10.1111/j.1747 -1796.2010.00400.x.

Vogel, Kathleen M. 2012. *Phantom Menace or Looming Danger? A New Framework for Assessing Bioweapons Threats*. Baltimore: Johns Hopkins University Press.

Wald, Priscilla. 2008. *Contagious: Cultures, Carriers, and the Outbreak Narrative*. Durham, NC: Duke University Press.

Wall Street Journal. 2001a. "The Anthrax Source." October 15, 2001.

———. 2001b. "The Anthrax War." October 18, 2001.

Warrick, Joby. 2002. "Missing Army Microbes Called Non-Infectious." *Washington Post*, January 22, 2002. https://www.washingtonpost.com /archive/politics/2002/01/22/missing-army-microbes-called-non -infectious/4422c125-a033-4eb9-a14b-2b2ba78516d1/.

Washington, Harriet A. 2012. "Biocolonialism." In *Deadly Monopolies: The Shocking Corporate Takeover of Life Itself—and the Consequences for Your Health and Our Medical Future*, 265–299. Reprint. New York: Anchor.

Watson, Crystal, Matthew Watson, Daniel Gastfriend, and Tara Kirk
Sell. 2018. "Federal Funding for Health Security in FY2019." *Health
Security* 16 (5). https://doi.org/10.1089/hs.2018.0077.

Weber, Stephen G., Ed Bottei, Richard Cook, and Michael O'Connor.
2004. "SARS, Emerging Infections, and Bioterrorism Preparedness."
Lancet Infectious Diseases 4 (8): 483–484. https://doi.org/10.1016/S1473
-3099(04)01098-9.

Weir, Lorna. 2014. "Inventing Global Health Security, 1994–2005." In
Routledge Handbook of Global Health Security, edited by Simon Rushton,
Jeremy R. Youde, and Jeremy R. Youde, 18–31. London: Routledge.

Weir, Lorna, and Eric Mykhalovskiy. 2010. "Emerging Infectious
Diseases: An Active Concept." In *Global Public Health Vigilance:
Creating a World on Alert*, 29–62. New York: Routledge.

Weiss, Rick, and Jo Warrick. 2002. "Army Lost Track of Anthrax
Bacteria." *Washington Post*, January 21. https://www.washingtonpost
.com/archive/politics/2002/01/21/army-lost-track-of-anthrax-bacteria
/09801da4-126c-4ed7-a685-18e555459cd3/.

Whidden, Michael. 2001. "Unequal Justice: Arabs in America and United
States Antiterrorism Legislation." *Fordham Law Review* 69 (January):
2825.

White, Lynn T. 2003. "SARS, Anti-Populism, and Elite Lies: Temporary
Disorders in China." In *The New Global Threat: Severe Acute Respiratory
Syndrome and Its Impacts*, edited by Tommy Thong Bee Koh, Aileen J.
Plant, and Eng Hin Lee, 31–67. Singapore: World Scientific.

White House, The. 2011. "Fact Sheet on the Successful Conclusion of the
Seventh Review Conference of the Biological and Toxin Weapons
Convention." Office of the Press Secretary. https://obamawhitehouse
.archives.gov/the-press-office/2011/12/23/fact-sheet-successful
-conclusion-seventh-review-conference-biological-an.

Whitt, Laurelyn. 2009. "Imperialism Then and Now." In *Science, Colonialism, and Indigenous Peoples: The Cultural Politics of Law and Knowledge*,
3–28. Cambridge, UK: Cambridge University Press.

WHO. *See* World Health Organization

Windrem, Robert. 2004. "The World's Deadliest Woman?" World
News, NBC News, September 23, 2004. http://www.nbcnews.com
/id/3340765/.

Wishnick, Elizabeth. 2010. "Dilemmas of Securitization and Health Risk Management in the People's Republic of China: The Cases of SARS and Avian Influenza." *Health Policy Plan* 25 (6): 454–466.

Wiwanitkit, Viroj. 2008. *Bird Flu: The New Emerging Infectious Disease.* Hauppauge, NY: Nova Science.

Workers Who Care: A Graphical Profile of the Frontline Health and Health Care Workforce. 2006. San Francisco: Health Workforce Solutions.

World Health Organization. n.d. "Cumulative Number of Confirmed Human Cases of Avian Influenza A/(H5N1) Reported to WHO." Accessed September 11, 2008. https://www.who.int/influenza/human_animal_interface/H5N1_cumulative_table_archives/en/.

———. 1999. "Smallpox Eradication: Destruction of Variola Virus Stocks." WHO. Fifty-Second World Health Assembly. A52/5. Provisional agenda item 13.

———. 2003a. "Severe Acute Respiratory Syndrome (SARS) Multi-country Outbreak—Update 2." March 17. http://www.who.int/csr/don/2003_03_17/en/.

———. 2003b. "Severe Acute Respiratory Syndrome (SARS) Multi-country Outbreak—Update 5." March 20. http://www.who.int/csr/don/2003_03_20/en/.

———. 2003c. "Summary of Probable SARS Cases with Onset of Illness from 1 November 2002 to 31 July 2003." December. http://www.who.int/csr/sars/country/table2004_04_21/en/.

———. 2003d. "Update 95—SARS: Chronology of a Serial Killer." July 4. https://www.who.int/csr/don/2003_07_04/en/.

———. 2004. *WHO Advisory Committee on Variola Virus Research: Report of the Sixth Meeting, Geneva, Switzerland, 4–5 November 2004.* http://www.who.int/csr/resources/publications/WHO_CDS_CSR_ARO_2005_4/en/.

———. 2005a. *Guidance for the Timely Sharing of Influenza Viruses/Specimens with Potential to Cause Human Influenza Pandemics.* March. http://flu.org.cn/en/news-7558.html.

———. 2005b. "Managing Biorisks in Laboratory Environments." February 3. https://www.who.int/csr/resources/publications/biosafety/meetingFeb_05/en/.

————. 2005c. "Revision of the International Health Regulations." WHA58.3. https://www.who.int/csr/ihr/WHA58-en.pdf.

————. 2005d. "WHO Checklist for Influenza Pandemic Preparedness Planning." http://www.who.int/csr/resources/publications/influenza /WHO_CDS_CSR_GIP_2005_4/en/.

————. 2005e. *WHO Global Influenza Preparedness Plan: The Role of WHO and Recommendations for National Measures before and during Pandemics.* Department of Communicable Disease Surveillance and Response Global Influenza Programme. https://www.who.int/csr/resources /publications/influenza/WHO_CDS_CSR_GIP_2005_5.pdf.

————. 2007. "Indonesia to Resume Sharing H5N1 Avian Influenza Virus Samples Following a WHO Meeting in Jakarta." March 27. http:// www.who.int/mediacentre/news/releases/2007/pr09/en/.

————. 2008. *Sixtieth World Health Assembly, Geneva, 14–23 May 2007: Summary Records of Committees: Reports of Committees.* https://apps.who .int/iris/handle/10665/22640.

————. 2010. "Pandemic (H1N1) 2009—Update 100." May 14. https:// www.who.int/csr/don/2010_05_14/en/.

————. 2011. *Pandemic Influenza Preparedness Framework: For the Sharing of Influenza Viruses and Access to Vaccines and Other Benefits.* https://apps .who.int/gb/pip/pdf_files/pandemic-influenza-preparedness-en.pdf.

————. 2012. "H5N1 Avian Influenza: Timeline of Major Events." https://www.who.int/influenza/human_animal_interface/H5N1_avian _influenza_update.pdf.

————. 2013. "Cumulative Number of Confirmed Cases for Avian Influenza A(H5N1) Reported to WHO, 2003–2013." http://www.who .int/influenza/human_animal_interface/EN_GIP_20130116Cumulativ eNumberH5N1cases.pdf?ua=1.

————. 2015. "Cumulative Number of Confirmed Cases for Avian Influenza A(H5N1) Reported to WHO, 2003–2015." http://www.who .int/influenza/human_animal_interface/EN_GIP_20150904cumulativ eNumberH5N1cases.pdf?ua=1.

Wortley, Pascale M., Benjamin Schwartz, Paul S. Levy, Linda M. Quick, Brian Evans, and Brian Burke. 2006. "Healthcare Workers Who Elected Not to Receive Smallpox Vaccination." *American Journal of*

Preventive Medicine 30 (3): 258–265. https://doi.org/10.1016/j.amepre
.2005.10.005.

Wright, Robin. 1995. "'Dr. Germ': One of the World's Most Dangerous
Women." *Los Angeles Times*, November 7, 1995.

Wright, Susan. 1989. "The Buildup That Was." *Bulletin of the Atomic
Scientists*, February.

———. 2002. "Double Language and Biological Warfare." *GeneWatch* 15
(April).

Wright, Susan, and Stuart Ketcham. 1990. "The Problem of Interpreting
the U.S. Biological Defense Research Program." In *Preventing A
Biological Arms Race*, edited by Susan Wright, 169–196. Boston: MIT
Press.

Yih, W. Katherine, Tracy A. Lieu, Virginia H. Rêgo, Megan A. O'Brien,
David K. Shay, Deborah S. Yokoe, and Richard Platt. 2003. "Attitudes
of Healthcare Workers in U.S. Hospitals Regarding Smallpox
Vaccination." *BMC Public Health* 3 (1): 20. https://doi.org/10.1186/1471
-2458-3-20.

Zamiska, Nicholas. 2006. "How Academic Flap Hurt World Effort on
Chinese Bird Flu." *Wall Street Journal*, February 24, 2006. http://www
.wsj.com/articles/SB114072620677781658.

Zamiska, Nicholas, and Marc Champion. 2006. "Global Bird-Flu
Preparedness Is Taking Shape." *Wall Street Journal*, February 16, 2006.

Index

Note: Page numbers in *italics* denote images/figures.

D'Harlingue, Benjamin, 31
discourse-making: as central to preparedness regimes, 17–19; and challenge to virus-sharing system, 128–129; and compliance enforcement, 119–120; counter-hegemonic discourse, 19, 99–100, 125–130; defined, 2–3; and vaccination campaigns, 90–94; and vulnerability narrative, 94–99. *See also* bioterror imaginary; media
discrimination against Arabs/Muslims, 32, 34, 67. *See also* bioterror imaginary; Orientalism; race and racialization
Dorman, David, 117
dual use, 66, 74–75

Ebright, Richard, 81
education, of Arab/Muslim scientists, 43, 44–45, 46
emerging infectious diseases/emerging diseases worldview, 109–110, 112, 170n12
Enloe, Cynthia, 98–99
E-Ring (TV show), 8
Eurocentrism, in pandemic preparedness, 109–111
exceptionalism, U.S., 5–6, 49–50, 57–58, 64–68, 111, 146n15
extractive biocolonialism (germs as resources), 120–124, 175nn37–39; challenge to, 125–130

Fauci, Anthony S., 87
Federal Bureau of Investigations (FBI), 29, 30, 71; and anthrax investigation, 1, 2, 144n4, 145n7
federal government, as research locus, 18
Federation of American Scientists (FAS), 79–80
femininity, and colonialism, 95–96, 155–156n29. *See also* gender; women
feminism: and critiques of imperialism/militarism, 23, 54–55, 101–103; in imperialist narratives, 5–6, 48, 49–50

Feminism and War: Confronting U.S. Imperialism (Riley, Mohanty, and Pratt), 58
feminist science studies, 20, 22, 150n43
feminized vulnerability. *See* vulnerability narrative
Fidler, David P., 87
flu. *See* influenza
flu samples: all countries required to provide to WHO, 113, 114; and H5N1 outbreak, 114–115; as resource, 120–124; restructuring of sharing system, 125–130
foreign nationals, 31, 34, 70, 134, 152n7, 159n25. *See also* immigrants/immigration; national origin
Foucault, Michel, 2, 17, 159n26
fragility, U.S. *See* vulnerability narrative
Fraser, Nancy, 17
freedom, discourse of: scientific freedom, 66, 79–80, 134, 160n31; in war on terror, 5, 12, 49–50
frontline health care workers: and feminized vulnerability, 101, 167n33; nurses' opposition to smallpox vaccination program, 84–85, 93, 99–100, 103, 138; racial composition of, 101, 167n34; in smallpox vaccination program, 83–84, 100

Garrett, Laurie, 129
gender: and colonialism, 95–96, 155–156n29; and depiction of Iraqi female scientists, 47–51; and feminized vulnerability, 101–102; masculinity, Arab/Muslim, 33–34, 34–35, 48; masculinity, white, 2–3, 76, 95–96, 145n8, 161n34; and national security discourse, 6–7; and war on terror, 6, 33–34, 36, 95–96; white women and vaccination program, 94–99. *See also* women
genetic engineering, 11, 72, 75, 77–78, 148–149n34
Geneva Protocol (1925), 27

pandemics: 1918 influenza, 75, 81, 168n2; emergence of, 108; H1N1, 108, 136–137, 174n33, 176n6
paternalism, U.S., 110–111, 131. *See also* U.S. exceptionalism
PATRIOT Act (2001), 31, 65, 70, 152n7
Perez-Pena, Richard, 96
Petersen, Melody, 7
pharmaceutical industry: and biodefense program, 63, 158n15; and vaccine production, 120–123, 124, 137, 158n15
Philipose, Liz, 55
"The Pitfalls of Bioterrorism Preparedness: The Anthrax and Smallpox Experiences" (Cohen, Gould, and Sidel), 80–81
Powell, Colin, 56–57
power, 145n9; activist organizations and social movements, 71, 134–136, 138–139; benefits sharing, 126–127, 137; biocolonialism, 120–124, 125–130, 175nn37–39; biopower, 71, 159–160n26; China/U.S. imbalance, 116; counter-hegemonic discourse, 19, 99–100, 125–130; and discourse, 2–3, 17–19; and international arms control, 55–56; and public health, 15, 102, 103 (*See also* public health, securitization of); and war on terror, 5, 46. *See also* vulnerability narrative
Prashad, Vijay, 116
preparedness, 8–9, 11, 12, 13–14, 104, 147n27. *See also* pandemic preparedness
Project Bacchus, 61
Puar, Jasbir, 6, 32, 36, 95, 153n14
public health, securitization of, 13–14, 24–25, 82–104; anti-racist, anti-imperialist critique of, 101–103; and bioterrorism preparedness, 13–16, 24–25; counternarratives to, 99–100; effects of, 103–105; and racialization of communities of color, 90–94; smallpox vaccination program, 83–87;

and white femininity, 94–99. *See also* pandemic preparedness
Public Health Security and Bioterrorism Preparedness and Response Act (Bioterrorism Preparedness Act, 2002), 10, 12, 72–73, 79

quarantines, 107, 108, 164n15, 169n7

race and racialization, 152n4, 164–165n19; of Arabs/Muslims, 4–5, 29–31, 35–36, 151n3, 153n12, 153n13, 159n25; of communities of color as public health threats, 90–94; in frontline health field, 101–102, 167n34; and national security discourse, 4–6, 146n15; as tool of social control, 145n12. *See also* Arabs and Muslims; bioterror imaginary; Orientalism
Rai, Amit, 32, 36, 95
Ramirez, Michael, 34–35
Rana, Junaid, 23, 153n14
Rasenberger, Jim, *88*
Reagan administration, 60
Riley, Robin L., 47, 58
Rosenberg, Barbara Hatch, 80
Royal Society (UK national academy of science), 53
Russia, 1, 60–61, 76, 78, 161n37. *See also* Soviet Union

Said, Edward, 29, 173n26
samples. *See* flu samples
Sarasin, Philipp, 40
SARS (severe acute respiratory syndrome) epidemic, 15, 106–107, 110–111, 116–117, 148n32, 171nn14–15, 172–173n25
science and scientists: attitudes toward biological warfare, 54, 79–80, 134, 160n31; and biodefense, 79–81, 134; movement against militarism in, 133–136; and progress narrative, 9, 37–38; and regime of biosecurity, 73–74. *See also* scientist-terrorist, figure of; technoscience, faith in

USA Today, 44
U.S. exceptionalism, 5–6, 49–50, 57–58,
    64–68, 146n15
U.S. fragility. *See* vulnerability
    narrative

vaccinations: complications of, 84, 93,
    96, 99, 165n21; historical, 88–89;
    National Smallpox Vaccination
    Program (NSVP), 82, 83–85, 86–87,
    91, 93–94, 99–100, 103, 104–105, 138,
    162n3
vaccine development, 120–121, 122–123,
    156–157n7
vaccine distribution, 126–127, 128,
    136–137, 174n33
vaccine production, 120–122, 126–127,
    128, 158n15
variola virus, 76–78, 161n38, 161n39.
    See also smallpox
victimhood narrative. *See* vulnerability
    narrative
violent male terrorist, figure of, 30, 33,
    34–38
virus-sharing system, 120–121, 122–123,
    176–177n7; Indonesia's challenge to,
    125–130, 137
vulnerability narrative, 147–148n28; and
    biosecurity discourse, 74, 80; and
    gender, 6, 95–99, 105; and global
    health governance, 128; and
    preparedness logic, 12–13; reframing
    of, and NSVP, 99–100, 101–103

Wall Street Journal, 1, 8, 33, 40
war on terror: and bio-imperialism,
    132–133, 134–135; discourse of, 4–7;
    discourse of civilization in, 5, 24, 28,
    32–33, 39–40, 65; discourse of
    freedom in, 5, 12, 49–50; and gender,
    6, 33–34, 36, 95–96; and media
    portrayals of Arabs/Muslims, 31–32;

and Orientalism, 5, 28–30; and
    WHO priorities, 112–113
Washington Post, 115, 129, 148n33, 148n34,
    150n41
weapons of mass destruction (WMDs),
    4, 112–113
white children, and public health/mass
    vaccinations, *98–99,* 166n29
white fragility, 95, 97, 147–148n28
white masculinity, 2–3, 95–96
white scientific masculinity, 76, 145n8,
    161n34
white women, and public health/mass
    vaccinations, 94–99, 101
WHO. *See* World Health Organization
"Why Revive a Deadly Flu Virus?"
    (NYT article), 81
women: Arab/Muslim, representations
    of, 6, 34, 47–51, 155n28; and bio-
    preparedness, 165n24; of color, in
    frontline health care work, 101–102,
    167n34; of color, representations of,
    155–156n29; in narratives of war on
    terror, 6; white, and public health/
    mass vaccinations, 94–99, 101
Wong, Alvin Ka Hin, 31
World Health Assembly (May 2007),
    127, 128
World Health Organization (WHO):
    and emerging diseases worldview,
    170n12; and H1N1 pandemic, 136;
    International Health Regulations
    (2005 IHR), 113–114, 120, 125,
    171–172n18, 173n30; and neocolonial-
    ism in pandemic preparedness,
    112–114, 120, 125; and SARS, 106,
    110–111, 171n15; and smallpox
    research, 76, 77, 78; and virus-sharing
    system, 120–121, 125–126, 127, 129
Wright, Susan, 60, 157n9

Youling, Jia, 114–115

About the Author

GWEN SHUNI D'ARCANGELIS is an activist scholar who studies the social dimensions of science and health. She is an associate professor of gender studies at Skidmore College, Saratoga Springs, New York. She has published on white scientific masculinity in U.S. security discourse, gendered Orientalism in media coverage of SARS, and nurse activism opposing the war on terror.